Studies in Computational Intelligence 435

Editor-in-Chief

For further volumes:
http://www.springer.com/series/7092

Takayuki Ito, Minjie Zhang, Valentin Robu,
and Tokuro Matsuo (Eds.)

Complex Automated Negotiations: Theories, Models, and Software Competitions

Springer

Editors
Takayuki Ito
Nagoya Institute of Technology
Nagoya
Japan

Valentin Robu
University of Southampton
Southampton
United Kingdom

Minjie Zhang
University of Wollongong
Wollongong
Australia

Tokuro Matsuo
Advanced Institute of
Industrial Technology
Tokyo
Japan

ISSN 1860-949X
e-ISSN 1860-9503
ISBN 978-3-642-30736-2
e-ISBN 978-3-642-30737-9
DOI 10.1007/978-3-642-30737-9
Springer Heidelberg New York Dordrecht London

Library of Congress Control Number: 2012940421

Printed on acid-free paper

Springer is part of Springer Science+Business Media (www.springer.com)

New Frontiers for Agent-Based Complex Automated Negotiations

Preface

Complex Automated Negotiations are a widely studied, emerging area in the field of Autonomous Agents and Multi-Agent Systems. In general, automated negotiations can be complex, since there are a lot of factors that characterize such negotiations. These factors include the number of issues, dependency between issues, representation of utility, negotiation protocol, negotiation form (bilateral or multi-party), time constraints, etc. Software agents can support automation or simulation of such complex negotiations on the behalf of their owners, and can provide them with adequate bargaining strategies. In many multi-issue bargaining settings, negotiation becomes more than a zero-sum game, so bargaining agents have an incentive to cooperate in order to achieve efficient win-win agreements. Also, in a complex negotiation, there could be multiple issues that are interdependent. Thus, agent's utility will become more complex than simple utility functions. Further, negotiation forms and protocols could be different between bilateral situations and multi-party situations. To realize such a complex automated negotiation, we have to incorporate advanced Artificial Intelligence technologies includes search, CSP, graphical utility models, Bays nets, auctions, utility graphs, predicting and learning methods. Applications could include e-commerce tools, decision-making support tools, negotiation support tools, collaboration tools, etc. For this book, we solicited papers on all aspects of such complex automated negotiations which are studied in the field of Autonomous Agents and Multi-Agent Systems.

This book includes two parts, which are Part I: Agent-based Complex Automated Negotiations and Part II: Automated Negotiation Agents Competition. Each chapter in Part I is an extended version of a ACAN 2011 paper after peer reviews by three PC members. Part II includes ANAC2011 (The Second Automated Negotiating Agents Competition), in which automated agents who have different negotiation strategies and implemented by different developers are automatically negotiate in the several negotiation domains. ANAC is an international competition in which automated negotiation strategies, submitted by a number of universities and research institutes across the world, are evaluated in a tournament style. The purpose of the competition is to steer the research in the area of bilateral multi-issue, closed negotiation. Closed negotiation, when opponents do not reveal their preferences to each other, is an important class of real-life negotiations. Negotiating agents designed using heuristic approach need extensive evaluation, typically through simulations and empirical analysis, since it is usually impossible to predict precisely how the system and the constituent agents will behave in a wide variety of circumstances, using purely theoretical tools. This book includes rules, results, participants agents and domains descriptions for ANAC2011 submitted by organizers and finalists. The reports from the ANAC2011 competition highlight important aspects, that should be considered in future works on automated negotiation.

Finally, we would like to extend our sincere thanks to all authors. This book would not have been possible without valuable supports and contributions of the cooperators. Especially, we would like to show our appreciation to *Katsuhide Fujita* from The University of Tokyo for his significant contributions to editing this book and *Raz Lin* from Bar-Ilan University for collecting the ANAC papers from the finalists.

Japan, April 1st, 2012

Takayuki Ito
Minjie Zhang
Valentin Robu
Tokuro Matsuo

Contents

List of Contributors

Enrique de la Hoz
Computer Engineering Department, Universidad de Alcala, Escuela Politecnica, 28871, Alcala de Henares (Madrid), Spain
e-mail: enrique.delahoz@uah.es

Miguel A. Lopez-Carmona
Computer Engineering Department, Universidad de Alcala, Escuela Politecnica, 28871, Alcala de Henares (Madrid), Spain
e-mail: miguelangel.lopez@uah.es

Ivan Marsa-Maestre
Computer Engineering Department, Universidad de Alcala, Escuela Politecnica, 28871, Alcala de Henares (Madrid), Spain
e-mail: ivan.marsa@uah.es

Mark Klein
Sloan School of Management, Massachusetts Institute of Technology, Cambridge, MA, U.S.A.
e-mail: m_klein@mit.edu

Katsuhide Fujita
School of Engineering, The University of Tokyo, Tokyo, Japan
e-mail: fujita@ipr-ctr.t.u-tokyo.ac.jp

Takayuki Ito
School of Techno-Business Administration, Nagoya Institute of Technology, Nagoya, Aichi, Japan
e-mail: ito.takayuki@nitech.ac.jp

Raiye Hailu
Department of Computer Science and Engineering, Nagoya Institute of Technology, Nagoya, Aichi, Japan
e-mail: raiye@itolab.nitech.ac.jp

Rafik Hadfi
School of Techno-Business Administration, Nagoya Institute of Technology, Nagoya, Aichi, Japan
e-mail: rafik@itolab.nitech.ac.jp

Chao Yu
School of Computer Science and Software Engineering, University of Wollongong, Australia
e-mail: cy496@uow.edu.au

Fenghui Ren
School of Computer Science and Software Engineering, University of Wollongong, Australia
e-mail: fren@uow.edu.au

Minjie Zhang
School of Computer Science and Software Engineering, University of Wollongong, Australia
e-mail: minjie@uow.edu.au

Tim Baarslag
Man Machine Interaction Group, Delft University of Technology, Delft, The Netherlands
e-mail: T.Baarslag@tudelft.nl

Koen Hindriks
Man Machine Interaction Group, Delft University of Technology, Delft, The Netherlands
e-mail: K.V.Hindriks@tudelft.nl

Catholijn Jonker
Man Machine Interaction Group, Delft University of Technology, Delft, The Netherlands
e-mail: C.M.Jonker@tudelft.nl

Reyhan Aydoğan
Man Machine Interaction Group, Delft University of Technology, Delft, The Netherlands
e-mail: R.Aydogan@tudelft.nl

Pınar Yolum
Department of Computer Engineering, Boğaziçi University, Bebek, 34342, Istanbul,Turkey
e-mail: pinar.yolum@boun.edu.tr

Mikoto Okumura
School of Techno-Business Administration, Nagoya Institute of Technology, Nagoya, Aichi, Japan
e-mail: okumura@itolab.nitech.ac.jp

Liviu Dan Şerban
Babeş-Bolyai University, Str. Theodor Mihali 58-60, 400591, Cluj-Napoca, Romania
e-mail: liviu.serban@econ.ubbcluj.ro

Cristina Maria Ştefanache
Babeş-Bolyai University, Str. Theodor Mihali 58-60, 400591, Cluj-Napoca, Romania
e-mail: cristina.stefanache@econ.ubbcluj.ro

Gheorghe Cosmin Silaghi
Babeş-Bolyai University, Str. Theodor Mihali 58-60, 400591, Cluj-Napoca, Romania
e-mail: gheorghe.silaghi,@econ.ubbcluj.ro

Cristian Marius Litan
Babeş-Bolyai University, Str. Theodor Mihali 58-60, 400591, Cluj-Napoca, Romania
e-mail: cristian.litan@econ.ubbcluj.ro

Paulo Maio
GECAD, School of Engineering, Polytechnic of Porto, Portugal
e-mail: pam@isep.ipp.pt

Nuno Silva
GECAD, School of Engineering, Polytechnic of Porto, Portugal
e-mail: nps@isep.ipp.pt

José Cardoso
University of Trás os Montes and Alto Douro, Vila Real, Portugal
e-mail: jcardoso@utad.pt

Raz Lin
Department of Computer Science, Bar-Ilan University, Ramat-Gan, Israel 52900
e-mail: linraz@cs.biu.ac.il

Sarit Kraus
Department of Computer Science, Bar-Ilan University, Ramat-Gan, Israel 52900, and Institute for Advanced Computer Studies, University of Maryland, College Park, MD 20742 USA
e-mail: sarit@cs.biu.ac.il

Asaf Frieder
Department of Computer Science, Bar-Ilan University, Ramat-Gan, Israel 52900
e-mail: asaffrr@gmail.com

Gal Miller
Department of Computer Science, Bar-Ilan University, Ramat-Gan, Israel 52900

Mai Ben Adar
Department of Computer Science, Bar-Ilan University, Ramat-Gan, Israel 52900
e-mail: nadavsof@gmail.com

Nadav Sofy
Department of Computer Science, Bar-Ilan University, Ramat-Gan, Israel 52900

Avshalom Elimelech
Department of Computer Science, Bar-Ilan University, Ramat-Gan, Israel 52900

Colin R. Williams
School of Electronics and Computer Science, University of Southampton, University Road, Southampton, SO17 1BJ
e-mail: crw104@ecs.soton.ac.uk

Valentin Robu
School of Electronics and Computer Science, University of Southampton, University Road, Southampton, SO17 1BJ
e-mail: vr2@ecs.soton.ac.uk

Enrico H. Gerding
School of Electronics and Computer Science, University of Southampton, University Road, Southampton, SO17 1BJ
e-mail: eg@ecs.soton.ac.uk

Nicholas R. Jennings
School of Electronics and Computer Science, University of Southampton, University Road, Southampton, SO17 1BJ
e-mail: nrj@ecs.soton.ac.uk

Radmila Fishel
Department of Information Systems Engineering, Ben-Gurion University, Israel
e-mail: rada.fishel@gmail.com

Maya Bercovitch
Department of Information Systems Engineering, Ben-Gurion University, Israel

Ya'akov (Kobi) Gal
Department of Information Systems Engineering, Ben-Gurion University, Israel

A.S.Y. Dirkzwager
Man-Machine Interaction Group, Delft University of Technology, Mekelweg 4, Delft, The Netherlands
e-mail: A.S.Y.Dirkzwager@student.tudelft.nl

M. Hendrikx
Man-Machine Interaction Group, Delft University of Technology, Mekelweg 4, Delft, The Netherlands
e-mail: M.J.C.Hendrikx@student.tudelft.nl

J.R. De Ruiter
Man-Machine Interaction Group, Delft University of Technology, Mekelweg 4, Delft, The Netherlands
e-mail: J.R.deRuiter@student.tudelft.nl

Thijs van Krimpen
Man Machine Interaction Group, Delft University of Technology, Delft, The Netherlands
e-mail: T.M.vanKrimpen@student.tudelft.nl

Daphne Looije
Man Machine Interaction Group, Delft University of Technology, Delft, The Netherlands
e-mail: D.Looije@student.tudelft.nl

Siamak Hajizadeh
Man Machine Interaction Group, Delft University of Technology, Delft, The Netherlands
e-mail: S.Hajizadeh@student.tudelft.nl

Shogo Kawaguchi
Department of Computer Science, Nagoya Institute of Technology, Nagoya, Aichi, Japan
e-mail: kawaguchi@itolab.nitech.ac.jp

Part I
Agent-Based Complex Automated Negotiations

Consortium Formation Using a Consensus Policy Based Negotiation Framework

Enrique de la Hoz, Miguel A. Lopez-Carmona,
Mark Klein, and Ivan Marsa-Maestre

Abstract. Although multiagent negotiation is usually seen as a process that had to seek the consensus of all the participants, there are situations where unanimous agreement either is not possible or simply the rules imposed by the system do not allow such unanimous agreement. One of this situations is consortium formation in brokerage events where a grand coalition is no viable and an optimal group partition is expected in order to maximize the probability of success of the different consortia. In this paper we propose a novel framework, consensus policy based mediation framework (CPMF), to be able to perform multiagent negotiations where the type of consensus by which an agreement meets in some specific manner the concerns of all the negotiators is considered as an integral part within the multiparty negotiation protocols. CPMF relies on a novel distributed agreement exploration protocol based on the optimization technique (GPS) and on the use of Ordered Weighted Averaging (OWA) operators for the aggregation of the agent preferences on the set of alternatives proposed by the mediator in each negotiation round. The mediation rules allow for a linguistic description of the type of agreements needed. A possible application of CPMF to a real-world scenario is shown. Experiments show that CPMF is able to manage negotiations efficiently following predefined consensus policies offering agreements for situations where quorum is not viable.

1 Introduction

Research on multiparty automated negotiation has focused on optimizing some type of social welfare measurement [5, 17, 4, 7, 12, 13]. Examples of such measurements

Enrique de la Hoz · Miguel A. Lopez-Carmona · Ivan Marsa-Maestre
Computer Engineering Department, Universidad de Alcala, Escuela Politecnica, 28871, Alcala de Henares (Madrid), Spain

Mark Klein
Center for Collective Intelligence, MIT Sloan School of Management, Massachusetts Institute of Technology

T. Ito et al. (Eds.): Complex Automated Negotiations, SCI 435, pp. 3–22.
springerlink.com

would be the *sum or product of utilities*, the *min* utility, etc. The main objective has been building efficient mechanisms and protocols to reach agreements among multiple participants while social welfare has been usually placed aside.

Some kind of social welfare criterion is incorporated within the search process in some works[2, 3, 10]. The approach chosen by these authors is is a mediated-approach that tries to build mechanisms to obtain fair agreements by using fair direction improvements in the joint exploration of the negotiation space. First, a solution is proposed by the mediator, then, agents provide their utility gradients in the solution, and finally the mediator proposes a new contract in the bisector or in an arbitrary direction which is considered fair enough. Some limitations can be pointed out, though. Firstly, they work only when utility functions are derivable and quasi-concave. Secondly, as long as the absolute value of the gradient is not considered, the marginal utility obtained by the agents in each negotiation round may not be fair. To conclude, even considering that the agents reveal also the gradient magnitude, the protocol is prone to untruthful revelations of information to bias the direction generated by the mediator.

To overcome the outlined limitations, we propose CPMF, a *Consensus Policy Based Mediation Framework for multi-agent negotiation*. CPMF relies on a novel distributed agreement exploration protocol based on the *Generalized Pattern Search* optimization technique (GPS) [9], and on the use of *Ordered Weighted Averaging* (OWA) operators [19]. This framework allows to search for agreements following predefined consensus policies, which may take the form of linguistic expressions in order to satisfy system requirements or to circumvent situations where unanimous agreements are not possible. We present a potential application of CPMF to brokerage events. A brokerage event is an event that gives the chance to meet consortium partners and drafting project proposals with the starting consortia to be able to get funding from public or private entities. We believe that the utilization of CPMF could guide the process of consortium building helping to identify the most potentially profitable consortium provided that participating entities are usually quite complex and dependant on the other participants. Also, the ability of expressing mediation rules in linguistic terms fit very well with the consortium-size restrictions that are usual in this kind of environments.

The rest of the paper is organized as follows. Next section presents first the GPS algorithm and the basic operation of the negotiation protocol. Section 3 focuses on the mechanisms used by the mediator to aggregate agents' preferences and Section 4 presents the agreement search process. Section 5 describes the application of CPMF to consortium formation.The last section summarizes our conclusions and sheds lights on some future research.

2 Protocol Description

We shall assume a set of n agents $A = \{A_i | i = 1, \ldots, n\}$ and a finite set of issues $X = \{x_l | l = 1, \ldots, s\}$, where each issue x_l can be normalized to a continuous or discrete range $d_l = [x_l^{min}, x_l^{max}]$. Accordingly, a *contract* is a vector $x' = \{x'_l = 1, \ldots, s\}$ defined by the issues values. Furthermore, we assume that each agent A_i has a real or virtual mapping $V_i : X \to \mathbb{R}$ function that associates with each contract x a value $V_i(x)$ that gives the payoff the agent assigns to a contract. The exact nature of this mapping needs not be known. All that we want to assume is that each agent has some means to formulate a preference function over a set of alternatives. Thus, the preference function can be described as a mapping function between the negotiation space contracts and the set of real numbers. We make a general assumption that the preference of each agent can be non-monotonic and non-differentiable. We only require the preferences to be rational.

The aim of the agents will be to reach an agreement on a contract x', maximizing their individual payoff and minimizing the revelation of private information.

Next, we describe in detail the GPS for unconstrained optimization, which is used in the construction of the negotiation protocol. GPS belongs to the family of *Direct Search Based* optimization algorithms [9]. Note, however, that our negotiation protocol is not a single-objective or multi-objective centralized optimization process.

2.1 Generalized Pattern Search Algorithm for Unconstrained Optimization

In this section we describe the distributed agreement exploration protocol used to identify potential contracts. At an iteration k of the protocol,we successively look at the points in the *mesh* $x^+(k) = x(k) \pm \triangle_k e_j$, $j \in \{1, \ldots, m\}$, where e_j is the jth standard basis vector and $\triangle_k > 0$ is a step-length parameter, to search for a contract $x'(k)$ in $x^+(k)$ for which $f(x'(k)) > f(x(k))$. We will denote the generated mesh (also called *pattern*)at round k as $x^{+o}(k)$ to designate the mesh. This mesh also includes the current point $x(k)$. We will use the notation $x^{e_j}(k) | j = 1, \ldots, 2m$ to describe each point in a mesh, and $x(k)$ or $x^{e_0}(k)$ to designate the current point. Fig. 1 illustrates the set of points among which we search for $m = 2$. If we find no $x'(k)$ such that $f(x'(k)) > f(x(k))$, then we reduce $\triangle_k$ by half and continue; otherwise, we leave the step-length parameter alone, setting $\triangle_{k+1} = \triangle_k$ and $x(k+1) = x'(k)$.

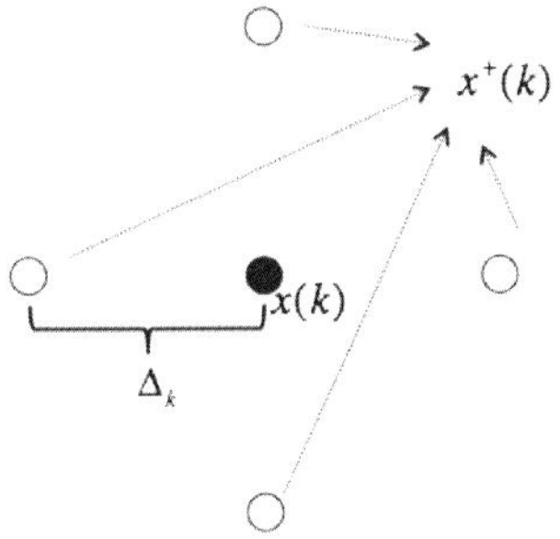

Fig. 1 An illustration of Generalized Pattern Search for unconstrained optimization

In the latter case we can also increase the step-length parameter, say, by a factor of 2, if we feel a longer step might be justified. We repeat the iteration just described until $\triangle_k$ is deemed sufficiently small. One important feature of pattern search that plays a significant role in a global convergence analysis is that we do not need to have an estimate of the derivative of f at $x(k)$ so long as the search includes a sufficient set of directions to form a positive spanning set for the cone of feasible directions, which in the unconstrained case is all of $\mathbb{R}^m$. In the unconstrained case the set $\{\pm e_j | j = 1, \ldots, m\}$ satisfies this condition, the purpose of which is to ensure that if the current iterate is not a stationary point of the problem, so that we have at least one ascendant direction.

The set e_j is defined by the number of independent variables in the objective function m and the positive standard basis set. Two commonly used positive basis sets in pattern search algorithms are the maximal basis, with $2m$ vectors, and the minimal basis, with $m + 1$ vectors.

2.2 *Basic Operation of the Negotiation Protocol*

The basic protocol of the proposed negotiation process is the following:

1. The mediator proposes a mesh from an initial contract $x^{ini}(1)$ for a step-length parameter $\triangle_1$. The point $x^{ini}(1)$ is randomly chosen by the mediator.
2. Each agent provides the mediator their preferences for the contracts in the current mesh x^{+o}, in terms of a mapping $S_i : X \rightarrow [0, 1]$ such that for example $S_i(x^{e_j}(k))$ indicates agent i's support for the alternative $x^{e_j}(k)$. An agent does not know the other agents' support for the contracts. Though agents are free to provide support values which are coincident or not with the corresponding private valuation function $V_i(x^{e_j}(k))$, in this work we will assume a perfect correspondence between both values.
3. The individual agent preferences for each contract are aggregated by the mediator to obtain the corresponding group preferences for each of the contracts in the mesh. We shall refer to this as the **aggregation of preferences** step.
4. The mediator decides which is the **preferred contract** in the mesh according to the group preferences for the different contracts.
5. Based on the **the preferred contract**, the mediator decides to **expand** or **contract** the mesh. Should a contraction make $\triangle_k$ small enough, the negotiation ends, otherwise go to step 2.

If we assume that a solution from X is always obtained, negotiation may end either when $\triangle_k$ is below a predefined threshold value or when a deadline expires. At each stage of the process an agent provides a support measure determined by its underlying payoff function and any information available about the previous stages of the negotiation, which constitutes a participating agent's strategy. The rules and procedures used in the negotiation process are crucial in an agent's determination of her strategy. In the following we shall describe the implementation of the negotiation process steps outlined above.

3 The Mediation Process

In this section we describe how the mediator aggregates the individual support for the contracts in the mesh at round k. Given a current contract $x(k)$ at round k, the starting point is a collection of n agents and a set $x^{+o}(k)$ of contracts (mesh). The mediator objective in this mediation step is to obtain a group preference function $G : x^{+o} \rightarrow [0,1]$ which associates with each alternative $x^{e_j}(k) \in x^{+o}(k)$ a value $G(x^{e_j}(k)) = M(S_1(x^{e_j}(k)), \ldots, S_n(x^{e_j}(k)))$, taking into account the preference $S_i(x^{+o}(k))$ expressed by every agent A_i over the set $x^{+o}(k)$. This preference indicates the degree to which each agent A_i supports each contract.

M is called the *mediation rule*, which describes the process of combining the individual preferences. The form of M can be used to reflect a desired mediation imperative or *consensus policy* for aggregating the preferences of the individual agents to get the mesh group preferences. M will guide the mediator in the expansion-contraction decisions in order to meet the desired type of agreements for the negotiation process.

We propose to use other mediation rules to improve the negotiation processes where either a quorum is not necessary or simply such quorum is not possible. For example, a solution may be acceptable if *most* of the agents support it. To incorporate these notions into our negotiation framework we will use a more general class of aggregation rules. The idea is to use a *quantifier guided aggregation*, which allows a natural language expression of the quantity of agents that need to agree on an acceptable solution. As we shall see the *Ordered Weighted Averaging* (OWA) operator [18] will provide a tool to model this kind of softer mediation rule.

3.1 Quantifier Guided Aggregation

Our aim is to define consensus policies in the form of a linguistic agenda for our mediation mechanisms. For example, the mediator should make decisions regarding the exploration of the negotiation space, i.e. expansion and contraction of the mesh, following mediation rules like "*Most* agents must be satisfied by the contract", "*at least* α agents must be satisfied by the contract","*many* agents must be satisfied", ...

The above statements are examples of *quantifier guided aggregations*. Zadeh [20] suggested a formal representation of these linguistic quantifiers using fuzzy sets. He suggested that any relative linguistic quantifier can be expressed as a fuzzy subset Q of the unit interval $I = [0,1]$. In this representation for any proportion $y \in I$, $Q(y)$ indicates the degree to which y satisfies the concept expressed by the term Q. In most applications of the quantifier guided aggregation we use a special case class of these linguistic quantifiers, called *Regular Increasing Monotone* (RIM) quantifiers. These types of quantifiers have the property that as more agents are satisfied our overall satisfaction can't decrease. Formally, these quantifiers are characterized in the following way: 1) $Q(0) = 0$, 2) $Q(1) = 1$ and 3) $Q(x) \geq Q(y)$ if $x > y$. Examples of this kind of quantifier are *all*, *most*, *many*, *at least* α. Two examples of RIM quantifiers are *all*, which is represented by Q_* where $Q_*(1) = 1$ and $Q_*(x) = 0$ for all $x \neq 1$, and *any*, which is defined as $Q^*(0) = 0$ and $Q^*(x) = 1$ for all $x \neq 0$.

One way to satisfy quantifies guided aggregation is by means of OWA operators An aggregation operator $M : S^n \rightarrow G, (S, G \in [0,1])$ is called an OWA operator of dimension n if it has an associated weighting vector $W = [w_1 w_2 \dots w_n]$ such that $w_t \in [0,1]$ and $\sum_{t=1}^n w_t = 1$ and where $M(S_1, \dots, S_n) = \sum_{t=1}^n w_t b_t$ where b_t is the tth largest element of the aggregates $\{S_1, \dots, S_n\}$.

In the definition of OWA we have used the notation M to identify the aggregation operator with the mediation rule, S^n to make reference to the preferences of the agents, and G to define the group preference. In the OWA aggregation the weights are not directly associated with a particular argument but with the ordered position of the arguments. If *ind* is an index function such that $ind(t)$ is the index of the tth largest argument, then we can express $M(S_1, \dots, S_n) = \sum_{t=1}^n w_t S_{ind(t)}$.

It can be shown the OWA operator is a mean operator. The form of the aggregation is dependent upon the associated weighting vector. We have a number of special cases of weighting vector which are worth noting. The vector W^* defined such that $w_1 = 1$ and $w_t = 0$ for all $t \neq 1$ gives us the aggregation $Max_i[S_i]$. Thus, it provides the largest possible aggregation. The vector W_* defined such that $w_n = 1$ and $w_t = 0$ for all $t \neq 1$ gives the aggregation $Min_i[S_i]$. The weighting vector W_{ave} defined such that $w_t = 1/n$ gives us the average $\frac{1}{n}\sum_{i=1}^n S_i$. Finally, an interesting family of OWA operators are the E-Z OWA operators. There are two families. In the first family we have $w_t = 1/q$ for $t = 1$ to q, and $w_t = 0$ for $t = q+1$ to n. Here we are taking the average of the q largest arguments. The other family defines $w_t = 0$ for $t = 1$ to q, and $w_t = \frac{1}{n-q}$ for $t = q+1$ to n. We can see that this operator can provide a softening of the original *min* and *max* mediation rules by modifying q.

The question now is how to obtain the OWA operator to satisfy a quantifier guided aggregation. Again, assume we have a collection of n agents. These agents have their preferences represented as fuzzy subsets over the set of alternatives in the mesh $\{S_1(x^{+o}(k)), \dots, S_n(x^{+o}(k))\}$. Under the quantifier guided mediation approach a group mediation protocol is expressed in terms of a linguistic quantifier Q indicating the proportion of agents whose agreement if necessary for a solution to be acceptable. The basic form of the mediation rule in this approach is that Q agents must be satisfied by the contract, where Q is a quantifier.

The formal procedure used to implement this mediation rule is described in the following:

1. Use Q to generate a set of OWA weights, $w_1, \dots, w_n$.
2. For each contract $x^{e_j}(k)$ in $x^{+o}(k)$ calculate the overall group support:

$$G(x^{e_j}(k)) = M(S_1(x^{e_j}(k)), \dots, S_n(x^{e_j}(k))).$$

The procedure used for generating the weights from the quantifier is to divide the unit interval into n equally spaced intervals and then to compute the length of the mapped intervals using Q

$$w_t = Q(\frac{t}{n}) - Q(\frac{t-1}{n}) \text{ for} t = 1, \dots, n.$$

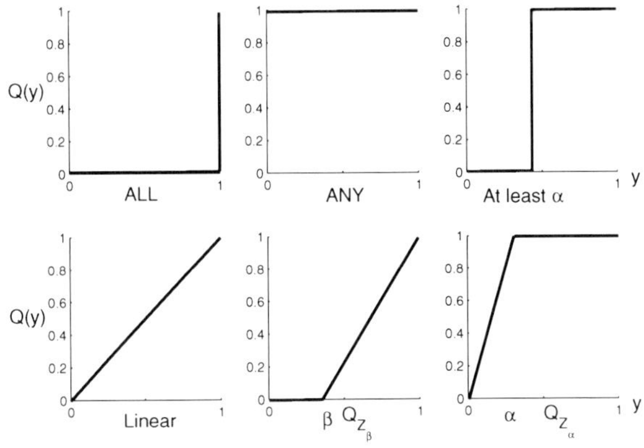

Fig. 2 Functional form of typical quantifiers: all, any, at least, linear, piecewise linear Q_{Z_β} and piecewise linear Q_{Z_α}

Because of the nondecreasing nature of Q it follows that $w_t \geq 0$. Furthermore from the regularity of Q, $Q(1) = 1$ and $Q(0) = 0$, it follows that $\sum_t w_t = 1$. Thus we can see that the weights generated are an acceptable class of OWA weights.

In Fig. 2 we show the functional form for the quantifiers *all*, *any*, Q_*, Q^*, *at least* α *percent*, *linear quantifier*, *piecewise* Q_{Z_β} and *piecewise* Q_{Z_α}. The quantifiers *all*, *any* and *at least* α describe the consensus policy using a natural language verbal description. However, more generally any function $Q : [0,1] \rightarrow [0,1]$ such that $Q(x) \geq Q(y)$ for $x \geq y$, $Q(1) = 1$ and $Q(0) = 0$ can be seen to be an appropriate form for generating mediation rules or consensus policies. Thus there are two techniques to generating these quantifier based mediation rules. One possibility is to start with a linguistic expression and then obtain Q. The second approach is to allow the mediation rule to be directly expressed in terms of a function Q. One important characteristic of this second method is that we can easily introduce into our mediation a number of formal properties that are not very easily expressed using a verbal description of the quantifier. The linear quantifier $Q(y) = y$ for instance generates $w_t = 1/n$, and thus, all the agents get the same weight. The Q_{Z_β} quantifier it is required that at least β agents are satisfied to initiate a Q linear improvement. Q_{Z_α} initiates the Q linear improvement with the first satisfied agent, and once there are α agents satisfied there is no improvement in Q if more agents are satisfied.

One feature which distinguishes the different types of mediation rules is the power of an individual agent to eliminate an alternative. For example, in the case of *all* this power is complete. In order to capture this idea the *Value Of Individual Disapproval* (VOID)

$$VOID(Q) = 1 - \int_0^1 Q(y)dy$$

measures this power. For the *all*, *any*, *at least* α and *linear* quantifiers the VOID measures are respectively 1, 0, α and 0.5. For the Q_{Z_β} quantifier $VOID(Q_{Z_\beta}) = \frac{1}{2} + \frac{\beta}{2})$ and therefore $VOID(Q_{Z_\beta}) \in [0.5, 1]$. The $Q_{Z\alpha}$ quantifier gets $VOID(Q_{Z\alpha}) = \frac{\alpha}{2}$ and $VOID(Q_{Z\alpha}) \in [0, 0.5]$.

Another family of quantifiers are those defined by $Q_p(y) = y^p$ for $p > 0$. In this case $VOID(Q_p) = 1 - \int_0^1 r^p dr = \frac{p}{p+1}$. For this quantifier we can easily obtain the OWA weights with

$$w_t = \left(\frac{t}{n}\right)^p - \left(\frac{t-1}{n}\right)^p.$$

For Q_p we see that as p increases we get closer to the *min* and that as p gets closer to zero we get the *max*.

3.1.1 Incorporating Agents' Relative Importance in Quantifier Guided Aggregation

Once we have introduced the quantifier guided aggregation, the mediation rule can be expressed in functional or linguistic form by statements of the type *Q agents are satisfied*. If we considered the consortium formation scenario, we have that entities can have different level of importance (e.g, regarding company size we can have large corporations and small and medium enterprises (SME)). To incorporate agents' relative importance we will use the Yager's approach [19] where the mediation rule is changed to *Q important agents must be satisfied*. In order to obtain the OWA weights we first define the importance I_i of an agent A_i as a non-negative real number, $I = \sum_1^n I_i$ as the sum of all the agents' importances, and $I_{ind(t)}$ the importance associated with the agent that gives the tth largest support to a contract. We make no restrictions on the total value of importances. To obtain these weights we use:

$$w_t = Q\left(\frac{\sum_{i=1}^{t} I_{ind(i)}}{I}\right) - Q\left(\frac{\sum_{i=1}^{t-1} I_{ind(i)}}{I}\right). \quad (1)$$

It is worth noting that the weights used in this aggregation will be different for each contract. This is due to the fact that the ordering of the support for the different contracts will usually change and in turn will lead to different values for each $I_{ind(i)}$.

Example. Let us assume that five agents $A_{i=1,...,5}$ are evaluating a contract, and that by means of some criterion the mediator has associated with these agents a set of importances $I_{i=1,...,5} = \{0.2, 0.5, 0.4, 2, 1\}$. The agents' support to the contract is given by $S_{i=1,...,5} = \{0.5, 0.2, 1, 0.1, 0.3\}$. We assume a $Q(y) = r^3$ *most* quantifier to guide the mediation. Table shows the ordered set of agent's supports and importances.

	S_i	I_i
A_3	1	0.4
A_1	0.5	0.2
A_5	0.3	1
A_2	0.2	0.5
A_4	0.1	2

The sum of importances is given by $I = \sum_{i=1}^{5} I_i = 4.1$. To compute the weights associated with the contract we use Eq. 1:

$$w_1 = Q\left(\frac{0.4}{4.1}\right) - Q\left(\frac{0}{4.1}\right) = \left(\frac{0.4}{4.1}\right)^3 - 0 = 9.28e-4 \quad w_2 = Q\left(\frac{0.6}{4.1}\right) - Q\left(\frac{0.4}{4.1}\right) = 0.0022$$
$$w_3 = Q\left(\frac{1.6}{4.1}\right) - Q\left(\frac{0.6}{4.1}\right) = 0.0563 \quad w_3 = Q\left(\frac{2.1}{4.1}\right) - Q\left(\frac{1.6}{4.1}\right) = 0.0749$$
$$w_5 = Q\left(\frac{4.1}{4.1}\right) - Q\left(\frac{2.1}{4.1}\right) = 0.8656$$

Thus, the group support to the contract is $G = M(0.5, 0.2, 1, 0.1, 0.3) = \sum_{t=1}^{5} w_t S_{ind(t)}$ $= 9.28e-4 \cdot 1 + 0.0022 \cdot 0.5 + 0.0563 \cdot 0.3 + 0.0749 \cdot 0.2 + 0.8656 \cdot 0.1 = 0.1205$

For the case of not considering the importance of the agents we obtain:

$$w_1 = Q\left(\tfrac{1}{5}\right) - Q\left(\tfrac{0}{5}\right) = 0.008 \quad w_2 = Q\left(\tfrac{2}{5}\right) - Q\left(\tfrac{1}{5}\right) = 0.056$$
$$w_3 = Q\left(\tfrac{3}{5}\right) - Q\left(\tfrac{2}{5}\right) = 0.152 \quad w_3 = Q\left(\tfrac{4}{5}\right) - Q\left(\tfrac{3}{5}\right) = 0.296$$
$$w_5 = Q\left(\tfrac{5}{5}\right) - Q\left(\tfrac{4}{5}\right) = 0.488$$

In this case the group support to the contract is $G = M(0.5, 0.2, 1, 0.1, 0.3) = \sum_{t=1}^{5} w_t S_{ind(t)} = 0.008 \cdot 1 + 0.056 \cdot 0.5 + 0.152 \cdot 0.3 + 0.296 \cdot 0.2 + 0.488 \cdot 0.1 = 0.1896$. The results show a smaller group support when considering the importance of the agents. This is an expected result provided that agents A_4, A_2 and A_5 concentrate most of the importance in the mediation process and they give a low support to the contract.

4 The Search Process

The search process is based on a mechanism whereby the mediator decides whether to generate a new mesh in order to continue with a new negotiation round, or to finish the negotiation process. This process starts just after any aggregation of preferences process, when the mediator has determined the group preferred contract $x^{e_*}(k)$. The relevant information available to the mediator at this point is at least the group preference $G(x^{+o}(k))$, the preferred contract $x^{e_*}(k)$, the current step-length $\triangle_k$, and the current round number k. With this information, the mediator has to select among three possible alternatives:

1. Move to the group preferred contract $x(k+1) = x^{e_*}(k)$ in $x^+(k)$ and expand the mesh by a factor of two $\triangle_{k+1} = 2 \cdot \triangle_k$.

2. Keep the current contract $x(k+1) = x(k)$ and reduce by half the mesh step-length $\triangle_{k+1} = \triangle_k/2$.
3. Finish the negotiation process.

For this paper we will assume what we call the *Standard Search Process* which selects among the mentioned alternatives as follows. The mediator selects alternative 1 if the preferred contract is in $x^+(k)$, i.e., $x^{e_*}(k) \in x^+(k)$. If the preferred contract is $x(k)$ then the mediator selects alternative 2. Finally, we define two stopping rules, one which bounds the maximum number of rounds k_{max}, and a second one which stops negotiation when the step-length $\triangle_k$ is below a predefined threshold γ. We assume that in both cases the agreement reached is the preferred group contract in the last negotiation round.

4.1 Contract Selection

In this section we describe the mechanisms used to select the preferred contract are described.The preferred contract selection process is a probabilistic process. The rationale behind using this probabilistic process is to introduce randomness and avoid local optima. With G the mediator is able to select a contract within the mesh. However, this selection is based on a relative measurement and it is not considering how good is the selection made. The mediator must consider both the G value and the relative values to make the decision of expansion and contraction

The point of departure is the set of final group preferences for the contracts in $x^{+o}(k)$ at round k. We propose a probabilistic selection process to select the winner contract in the mesh at a round k. We associate with each contract $x^{e_j}(k) \in x^{+o}(k)$ a probability

$$P(x^{e_j}(k)) = \frac{G(x^{e_j}(k))^{\sigma}}{\sum_j G(x^{e_j}(k))^{\sigma}}.$$

The process selects the winner contract using a biased random experiment with these probabilities. The parameter $\sigma > 0$ works as an indication of the significance we give to the final group preferences. If $\sigma \to \infty$ we select the contract with the maximum support, which means that the mediator is given the higher significance to the group preferences. If $\sigma = 1$ then the probability of selecting $x^{e_j}(x)$ would be proportional to its group support.

To introduce randomness and avoid local optima, we make σ vary as a function of G and the number of rounds k. If G is high, σ must be high, favouring a deterministic mesh movement, i.e. with a high probability the contract with a higher G is selected. Otherwise, if G is low, σ must be low to induce randomness and avoid local optima. More specifically, for $\sigma = 0$ the selection of contracts is equiprobable, making such selection independent of G. For $\sigma = 1$ the selection probability is proportional to G. Higher values for σ increases the probability of choosing the contract with a higher G. To control σ we define

$$\sigma(k,G) = \sigma_{min} + (\sigma_{max} - \sigma_{min}) \cdot G^{(1-\frac{k}{k_{max}})\cdot\alpha},$$

where σ depends on the negotiation round k, the maximum number of rounds k_{max} and G. The function is bounded by σ_{max} and σ_{min} given $G = 0$ and $G = 1$ respectively. The parameter $\alpha > 0$ determines the curvature of $\sigma(k,G)$. As the number of rounds k increases, the function increases its concaveness, which means that G induces higher values for σ, favouring convergence. The principle of this approach is analogous to the simulated annealing technique [5] without reannealing. We can also introduce reannealing for $k_r < k_{max}$ such that k/k_{max} converts into $\frac{k-k_r}{k_{max}-k_r}$.

5 Experimental Evaluation

In this section, we test our negotiation framework and show that the mechanisms proposed provide the mediator the tools to efficiently conduct multiagent negotiations by considering different consensus policies.

In the experimental setup, without loosing generality, we have considered 7 agents, 2 issues and 2 different types of negotiation spaces: a negotiation space where agents' utility functions are strategically built to define a *proof of concept negotiation scenario*, and a *complex negotiation scenario* where utility functions exhibit a more complex structure. In both cases utility functions are built using an aggregation of *Bell functions*. This type of utility functions capture the intuition that agents' utilities for a contract usually decline gradually with distance from their ideal contract. Bell functions are ideally suited to model, for instance, spatial and temporal preferences [14, 11]. In addition, they allow to configure different negotiation scenarios in terms of different complexity degrees.

In the *proof of concept negotiation scenario* each agent has a utility function with a single optimum. Fig. 3 shows in the same graph the agents' utility functions in the bidimensional negotiation space $[0,100]^2$. In this scenario four agents (Agent 1, 2, 3, 4) are in weak opposition (i.e. their preferences are quite similar), Agents 6 and 7 are in weak opposition and in very strong opposition with respect the other agents, and Agent 5 is in very strong opposition with respect the rest of the agents. In the *complex negotiation scenario* each agent's utility function is generated using two randomly located bells. The radius and height of each bell are randomly distributed within the ranges $r_i \in [20,35]$ and $h_i = [0.1,1]$. Fig. 4 shows the utility functions generated for each agent in this second case.

The configuration of parameters in the mediator is: $k_{max} = 50$ rounds, mesh tolerance $1e-6$, and $\alpha = 2$, $\sigma_{min} = 1$, $\sigma_{max} = 200$ for the preferred contract selection process. Previous experiments have confirmed that these parameter values perform well under most negotiation scenarios.

We tested the performance of the protocol under the proof of concept and complex negotiation scenarios for 5 different consensus policies defined by the corresponding VOID degrees: 0, 0.25, 0.5, 0.75 and 0.95, using the quantifier $Q_p(y) = y^p$. We also define a contrast experiment where the consensus policy based mediation

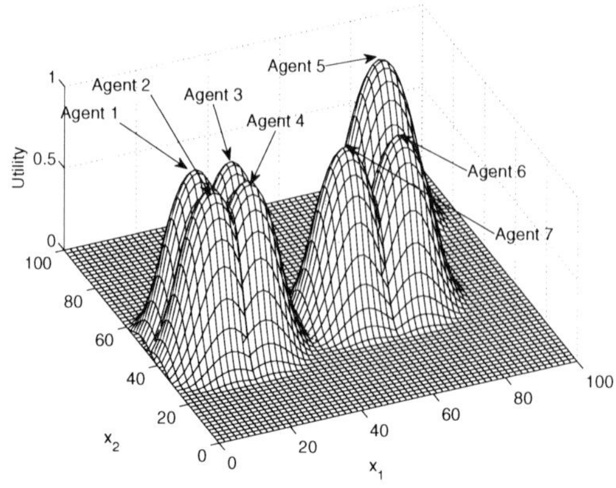

Fig. 3 Utility functions for the *proof of concept negotiation scenario*

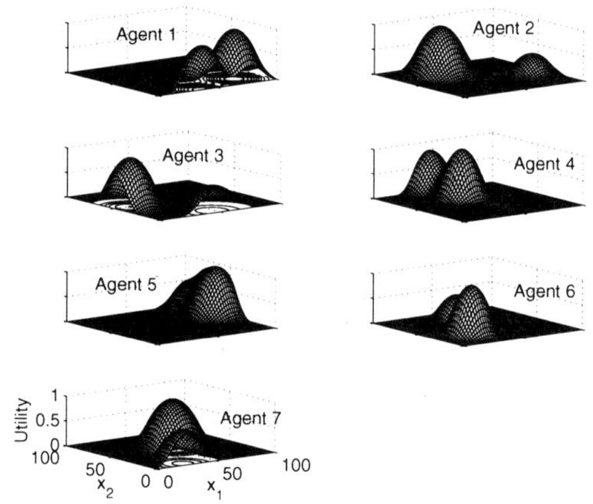

Fig. 4 Utility functions for the *complex negotiation scenario*

process is deactivated, such that the mediator uses the pattern search based process but there is no randomness and the group preference evaluation is limited to compute the sum of agents' valuations for a given contract (i.e. the winner contract is that with the highest sum of valuations).This experiment uses also 50 rounds and a mesh tolerance $1e-6$.

Each experiment consists of 100 negotiations where we capture the utilities achieved by each agent. First, we build a $7\,\text{agents} \times 100\,\text{negotiations}$ utility matrix where each row provides each agent's utilities and each column is a negotiation. Then, the matrix is reorganized such that each column is individually sorted from higher to lower utility values. Note that after this transformation the association row/particular-agent disappears. Given the matrix, we form 7 different utility groups: a first group named *group level 1* where we take the highest utility from each

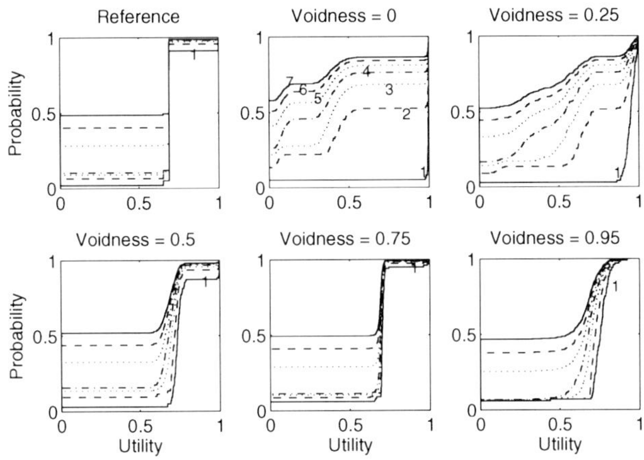

Fig. 5 Cumulative distributions of utilities for the *proof of concept scenario*

negotiation (i.e. the first row), a second group named *group level 2* with the two first rows and so on. In order to show the performance of the protocol we have used the Kaplan-Meier estimate of the cumulative distribution function (*cdf*) of agents' utilities for each group. Thus, we compute the *cdf* for the highest utilities, for the two highest utilities and so on. The *cdf* estimates the probability of finding agent's utilities below a certain value. The rationale behind using grouping in the analysis is to evaluate the ability of the protocol to find solutions which satisfy groups of agents.

In the proof of concept scenario (see Fig. 3) it can be seen that when a quorum is needed, the best alternative is to get satisfied agents 1, 2, 3 and 4. If it is enough to have one agent satisfied, any of the utility peaks would be a good solution. In Fig. 5 we show the results for the proof of concept scenario. Each line shows the *cdf* for a group level and the number above each line identifies the corresponding level. For instance, for the reference experiment and the group level 1 there is approximately a 98% probability of having agents with a utility 0.7, and a 2% probability of having agents with utility 0. In the group level 7 case, there is a 50% probability of having agents with utility 0.7, and a 50% probability of having agents with utility 0. For a VOID=0 and group level 1, however, the probability of having agents with a utility 1 is around 98%, which means that the mediator is applying efficiently the consensus policy which states that it is good enough to have one agent satisfied. As VOID increases (i.e. as it is necessary to have more agents satisfied) the *cdf* for group level 1 performs worse, though better than in the reference scenario, and for higher group levels the performance increases.

In Fig. 6 are shown the results for the complex negotiation scenario. Here we can also see how as VOID increases, the mediator biases the search for agreements where more agents are satisfied at the expense of not having individual agents highly satisfied. Globally, the results show that the proposed mechanisms are able to focus

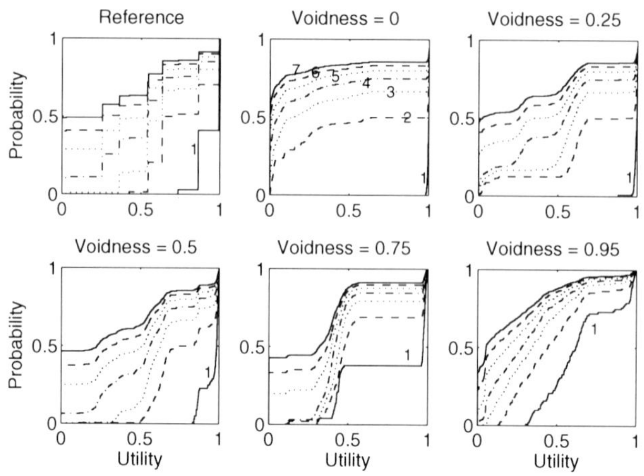

Fig. 6 Cumulative distributions of utilities for the *complex negotiation scenario*

the negotiation process in terms of consensus policies and to obtain better results than when using a classical welfare maximization approach.

6 A Proposal for the Application of CPMF to Consortium Formation

6.1 Problem Description

It is a common practice for the creation of new initiatives in the fields of joint R&D and business collaboration, the celebration of brokerage events with special emphasis in the development of new ideas for large and small scale public or private funded research projects in the field of applied industrial R&D to help with the preparation of drafting proposals and search for the right consortium partners.

A brokerage event is an event that gives the chance to meet consortium partners and drafting project proposals with the starting consortia. During these meetings, interested companies, research institutes, universities etc. with innovative ideas for projects are brought together which makes it a unique starting point for organizations to form successful consortia and start preparing project outlines. Participants propose their ideas to potential project partners, check project ideas from others to join their consortium, work out project ideas to indicative proposal and, finally, build a project consortium. These events serve also as an important indication for public authorities to sense the field of interest in advance.

It has been pointed out that not all participants get a good overview on all the information available on cooking project ideas. To improve the availability and access to project idea information and to maximize the opportunities for consortia building, participants should make their project ideas known through a web tool. Companies are able to register online to participate in brokerage events and, once registered, they are able to submit a full profile of their business. Partners that have uploaded their ideas in will get the opportunity to present their idea and to get a reserved meeting space.

We propose to extend this tool to support consortium building. Public authorities that promote these events are concerned with providing the ground for the establishment of international R&D networks involving companies, suppliers, R&D organizations and customers, from which strong consortia could be promoted in order to maximize the probability of successful proposals for market oriented joint contract research and development projects. That is why both companies and public authorities are interested in the formation of well-balanced and potentially successful consortia.

This tool would require each partner to prepare an outline of the type of partner they are looking for. This can be formalized with the partner preference matrix that we present later. Also, a company catalogue is issued to participants for review once all company profiles are received. We envision a mediated process where participants are able to express preference for companies they would like to be matched. Event organisers then take care of matching partners taking into account the project call rules. Moreover, there could be requirements about the size or nature of the members of the consortium (minimum or maximum amount of R&D organizations or industry involved). For instance, in the ITEA2 programme[1] promoted by EU projects must involve between 40 and 300 person-years.Taking into account partner preferences, the mediator would provide a feedback proposal to the partners, and the process would repeat until an agreement is found or a deadline is reached. The outcome of this negotiation process is not be seen as a hard constraint, but as a series of suggestions about potential consortium sets that maximize the funding raised by the participant entities.

6.2 Related Work

Coalition formation is an important problem in multi-agent systems, where several agents must be partitioned into teams so that their global utility is maximized. Agents can easily be used to represent the negotiating entities, and they are capable of finding more profitable coalitions than humans in complex environments. The notion of forming coalitions is normally applied in various domains, including commerce, sensor networks and social planning [16, 1].

When considering coalition formation we have to take into account two type of agents: cooperative agents (usually in a scenario of full information sharing) and self-interested agents. This second basic type of agents is a kind of agents that do not share their goals. Each self-interested agent has its own individual utility function, which is trying to maximize. Even such self-interested agents may still need to

[1] http://www.itea2.org/

cooperatively coordinate and collaborate with each other in many situations, though. One class of examples is consortium formation: individually, an agent cannot be eligible for any funding but if this agent forms a coalition with one or more other agents, the combined resources and joint effort of all agents in such a coalition may provide utility benefits to everyone involved.

There are situations where agents can cooperate by forming one coalition in which all the agents are members. This is called grand coalition. However, there are environments where the cost of cooperation is an ascending function in the number of cooperating agents where the partition of agents into subgroups would reduce their costs (since coordination and communication come about only within subgroups). This kind of environments is called non super-additive environments. Consortium formation is one of this environments because of both the overhead of communication in big groups and also due to the rules governing this fund-raising processes, which usually set a maximum in the number of partners that can apply for one of this grants.

The coalition formation process can generally be considered to include three main activities. First, the value of every possible coalition that can be formed is computed. Second, there is computation of the set of disjoint coalitions that have the maximum total value.Finally, a payoff distribution is to be determined to show the rewards that each agent in a coalition should obtain as a result of the actions taken by the coalition as a whole.

Coalition formation algorithms try to determine which of the potential coalitions should actually be formed and then they calculate a value for each coalition, known as the coalition value. This value provides an indication of the expected outcome derived from every coalition. Most of the proposals in the literature assume that this calculation is quite straightforward provided that utility functions are linear. Nevertheless, there are cases, as the one that we are dealing with, where this assumption does not hold because of the complexity of agents' utility function. Therefore, it is necessary to propose new ways of dealing with the process of evaluating the goodness of coalitions. Our work focuses on computing those potential coalitions and their corresponding coalition value.

The pursued goal is to find a combination of coalitions in which every agent belongs to exactly one coalition, and by which the overall outcome of the system is maximized. This problem is known as the coalition structure generation problem [8] and it is extremely challenging due to the number of possible combinations which grows very quickly as the number of agents increases, making it impossible to go through the entire search space, even for small numbers of agents . It is usually assumed that the value of a coalition does not depend on the actions of non members, which means that every coalition performs equally well, given any coalition structure containing it. These settings are known as characteristic function games (CFGs) [6], where the value of a coalition is given by a characteristic function.

Under these settings, the coalition structure generation problem can be seen as a complete set partitioning problem; given a collection of subsets of a ground set, and given a weight associated to each of these subsets, the set partitioning problem is to find an optimal way to partition the ground set. Complete means that every possible subset is included in the input, every possible coalition is taken into consideration. Unfortunately, it has been shown [15] that finding an optimal solution is NP-complete due to the fact that the space of possible solutions grows very rapidly with the number of elements involved.

We propose a method that uses CPMF to build consortia in the brokerage event scenario described in the previous section, where self-interested agents helped by a mediator can identify potential partners in order to team up for requesting funding for R&D projects. The final outcome consists of the mediator suggesting a series of potential consortia taking into account not only agent's preferences but also any rules imposed by the funding entity (e.g., European Union) regarding consortium size and member nature (large industries, small and medium sized enterprises and universities or research institutes).

6.3 *Consortium Formation Protocol*

We shall assume a set of n agents $A = \{A_i | i = 1,\dots,n\}$ representing the entities willing to be part of a consortium in order to prepare a project proposal. A consortium C is a non-empty subset of A. The main goal of the protocol is to find a non-overlapping consortium structure CS such that the potential amount of money raised (sum of the potential money raised by every partner) is maximized. A non-overlapping coalition structure is a set of coalitions, $CS = \{C_1,\dots,C_k\}$ such that the following condition is accomplished:

$$C_i \bigcap C_j = \emptyset \forall i, j = 1,\dots k, i \neq j$$

We denote the set of all possible partitions of the agent set A as $S(A)$. There will be a minimum and a maximum amount of money that can be requested by a project proposal, let say F_{min} and F_{max}. The amount of money will be distributed among the participating entities taking into account the degree of participation of each entity ('effort') in the project. An agent A_i will try to maximize the amount of money she gets from the project or, which is the same, her effort as they are proportional. There are some constraints, though, imposed by the resources each company can devote to the project so that there will be a maximum that in the most general case will be different for each of the entities.

Under this settings, a *contract* is a vector $x' = \{x'_l = 1,\dots,n \mid F_{min} \leq \sum_{l=1}^{n} x'_l \leq F_{max}\}$ defined by the issues values where x_l stands for the participation of agent A_l in the potential project funding. Accordingly, $F = \sum_{l=1}^{n} x'_l$. A contract x' implies a coalition $C_{x'}$ defined by:

$$C_{x'} = \{A_i \in A \mid x'_i \neq 0\}$$

We assume that each agent A_i has a mapping $V_i : X \rightarrow \mathbb{R}$ function that associates with each contract x a value $V_i(x)$ that gives the payoff the agent assigns to a contract. In this environment, each agent will associate each contract with the expected amount of funding she gets from that. It has to be noticed that as long as they are teaming up for a project proposal, there is no actual funding unless the project proposal is selected. This is the reason why it has to be taken into account not only agent's effort (x'_l) but also the probabilities of the project being successful as seen by the agent itself. To compute this probabilities, Agent A_i maintains a n-by-n preference matrix W where element w_{ij} means the probability of the consortium to succeed in getting the funding for the project if both A_i and A_j are in the consortium.

Definition 0.1. The *success probability* of the consortium $C_{x'}$ as seen by agent A_i, $P^i_{C_{x'}}$, is defined as:.

$$P_i(C_{x'}) = W_i x'$$

The *utility function* of agent A_i,V_i, is defined as the funding that agent A_i expects to get (x'_i) weighted by the probability $P_i(C_{x'})$:

$$V_i(x') = P_i(C_{x'})x'_i$$

Agents will try to reach an agreement in a contract x' that gives the amount of funding that would be obtained by each agent provided the consortium succeed in getting the funding. CPMF will be used to try to identify an agreement. Mediator will guide the search process a consensus rule that will translate the funding program rules usually expressed in terms of maximum and minimum of agents involved. It is important to note that a grand coalition is not possible here, that is why it is important to find solutions, in the form of potential consortia, the satisfy only a subset of agents. OWA operators make it possible to apply this linguistic rules into the search process.

Mediator will conduct a number of negotiations that will make possible to identify a set of potential coalitions that maximize the funding raised by the partners. As a result, we have a set of potential overlapping consortia OCS that comply with the size-constraint requirements while maximizing the utility obtained by agents. The size constraint is enforced by means of using the proper OWA. Notice that this is not a partition set but a number of high-valued potential consortia.

$$OCS = \{C_1, \ldots, C_k \mid C_i \bigcap C_j \neq \emptyset\}$$

After that, in a second phase, a process will be performed to identify a subset of non-overlapping coalitions from the high-valued consortia contained in OCS. This subset will be the coalition structure CS_{opt} that maximizes the utility distribution.

7 Conclusion

The consensus type by which an agreement meets in some specific manner the concerns of all the negotiators should be considered as in integral part of multiparty negotiation protocols. Nevertheless, most of the works focus only on unanimous agreements which do not fit well on every environment. Moreover, there situations where this kind of consensus is not desirable as the consortium formation scenario in brokerage events. We propose a multiagent negotiation protocol where the mediation rules at the mediator may take the form of a linguistic description of the type of consensus needed using OWA operator and quantifier guided aggregation.This protocol uses mechanisms derived from the Generalized Pattern Search non-linear optimization technique for the distributed exploration of the contract space in a process governed by the mediator that performs an aggregation of the agent preferences in each negotiation round applying the type of consensus desired.Experimental results show that CPMF applies efficiently predefined consensus policies and solves situations where unanimous agreements are not viable.

We believe that the negotiation framework presented opens the door to a new set of negotiation algorithms where consensus criteria will play an important role. One possible scenarios for this algorithms is consortium building in brokerage events where the linguistic expressed mediation rules could be of great utility for guiding the set partitioning process and the identification of high-valued consortia as it has been presented. Further research on the consortium formation is still necessary to test it under different assumptions. Also, the strategy issue remains opened and mechanisms need to be implemented to avoid or mitigate the incentive compatibility problem.

Acknowledgements. This work has been supported by the Spanish Ministry of Education and Science grants: TIN2008-06739-C04-04, T2C2 research project, and the mobility grant JC2010-003.

References

1. Dang, V.D., Dash, R.K., Rogers, A., Jennings, N.R.: Overlapping coalition formation for efficient data fusion in multi-sensor networks. In: Proceedings of the 21st National Conference on Artificial Intelligence, AAAI 2006, vol. 1, pp. 635–640. AAAI Press (2006)
2. Ehtamo, H., Hamalainen, R.P., Heiskanen, P., Teich, J., Verkama, M., Zionts, S.: Generating pareto solutions in a two-party setting: constraint proposal methods. Management Science 45(12), 1697–1709 (1999)
3. Heiskanen, P., Ehtamo, H., Hamalainen, R.P.: Constraint proposal method for computing pareto solutions in multi-party negotiations. European Journal of Operational Research 133(1), 44–61 (2001)
4. Ito, T., Klein, M., Hattori, H.: A multi-issue negotiation protocol among agents with nonlinear utility functions. Journal of Multiagent and Grid Systems 4(1), 67–83 (2008)
5. Klein, M., Faratin, P., Sayama, H., Bar-Yam, Y.: Protocols for negotiating complex contracts. IEEE Intelligent Systems 18(6), 32–38 (2003)
6. Kohlberg, E.: On the nucleolus of a characteristic function game. SIAM Journal on Applied Mathematics 20(1), 62–66 (1971), `http://www.jstor.org/stable/2099885`
7. Lai, G., Sycara, K.: A generic framework for automated multi-attribute negotiation. Group Decision and Negotiation 18, 169–187 (2009)

8. Lau, H.C., Zhang, L.: Task allocation via multi-agent coalition formation: Taxonomy, algorithms and complexity. In: Proceedings of the 15th IEEE International Conference on Tools with Artificial Intelligence, ICTAI 2003, p. 346. IEEE Computer Society, Washington, DC (2003), `http://dl.acm.org/citation.cfm?id=951951.952314`
9. Lewis, R.M., Torczon, V., Trosset, M.W.: Direct search methods: then and now. Journal of Computational and Applied Mathematics 124, 191–207 (2000)
10. Li, M., Vo, Q.B., Kowalczyk, R.: Searching for fair joint gains in agent-based negotiation. In: Decker, Sichman, Sierra, Castelfranchi (eds.) Proc. of 8th Int. Conf. on Autonomous Agents and Multiagent Systems, AAMAS 2009, Budapest, Hungary, pp. 1049–1056 (2009)
11. Lopez-Carmona, M.A., Marsa-Maestre, I., De La Hoz, E., Velasco, J.R.: A region-based multi-issue negotiation protocol for nonmonotonic utility spaces. Computational Intelligence 27(2), 166–217 (2011), doi:10.1111/j.1467-8640.2011.00377.x
12. Lopez-Carmona, M.A., Marsa-Maestre, I., Ibanez, G., Carral, J.A., Velasco, J.R.: Improving trade-offs in automated bilateral negotiations for expressive and inexpressive scenarios. Journal of Intelligent & Fuzzy Systems 21, 165–174 (2010)
13. Lopez-Carmona, M.A., Marsa-Maestre, I., Klein, M., Ito, T.: Addressing stability issues in mediated complex contract negotiations for constraint-based, non-monotonic utility spaces. Journal of Autonomous Agents and Multiagent Systems, 1–51 (2010) (published online)
14. Marsa-Maestre, I., Lopez-Carmona, M.A., Velasco, J.R., Ito, T., Klein, M., Fujita, K.: Balancing utility and deal probability for auction-based negotiations in highly nonlinear utility spaces. In: 21st International Joint Conference on Artificial Intelligence, IJCAI 2009, Pasadena, California, USA, pp. 214–219 (2009)
15. Sandholm, T., Larson, K., Andersson, M., Shehory, O., Tohmé, F.: Anytime coalition structure generation with worst case guarantees. In: Proceedings of the Fifteenth National/Tenth Conference on Artificial Intelligence/Innovative Applications of Artificial Intelligence, AAAI 1998/IAAI 1998, pp. 46–53. American Association for Artificial Intelligence, Menlo Park (1998)
16. Tsvetovat, M., Sycara, K., Chen, Y., Ying, J.: Customer Coalitions in Electronic Markets. In: Dignum, F., Cortés, U. (eds.) AMEC 2000. LNCS (LNAI), vol. 2003, pp. 121–138. Springer, Heidelberg (2001)
17. Vo, Q.B., Padgham, L., Cavedon, L.: Negotiating flexible agreements by combining distributive and integrative negotiation. Intelligent Decision Technologies 1(1-2), 33–47 (2007)
18. Yager, R.: Quantifier guided aggregation using owa operators. International Journal of Intelligent Systems 11, 49–73 (1996)
19. Yager, R., Kacprzyk, J.: The Ordered Weighted Averaging Operators: Theory and Applications. Kluwer (1997)
20. Zadeh, L.: A computational approach to fuzzy quantifiers in natural languages. Computing and Mathematics with Applications 9, 149–184 (1983)

The Effect of Grouping Issues in Multiple Interdependent Issues Negotiation between Exaggerator Agents

Katsuhide Fujita, Takayuki Ito, and Mark Klein

Abstract. Most real-world negotiation involves multiple interdependent issues, which makes an agent's utility functions complex. Traditional negotiation mechanisms, which were designed for linear utilities, do not fare well in nonlinear contexts. One of the main challenges in developing effective nonlinear negotiation protocols is scalability; it can be extremely difficult to find high-quality solutions when there are many issues, due to computational intractability. One reasonable approach to reducing computational cost, while maintaining good quality outcomes, is to decompose the contract space into several largely independent sub-spaces. In this paper, we propose a method for decomposing a contract space into sub-spaces based on the agent's utility functions. A mediator finds sub-contracts in each sub-space based on votes from the agents, and combines the sub-contracts to produce the final agreement. We demonstrate, experimentally, that our protocol allows high-optimality outcomes with greater scalability than previous efforts. Any voting scheme introduces the potential for strategic non-truthful voting by the agents, and our method is no exception. For example, one of the agents may always vote truthfully, while the other exaggerates so that its votes are always "strong." It has been shown that this biases the negotiation outcomes to favor the exaggerator, at the cost of reduced social welfare. We employ the limitation of strong votes to the method of decomposing the contract space into several largely independent sub-spaces. We investigate whether and how this approach can be applied to the method of decomposing a contract space.

Katsuhide Fujita
Institute of Engineering Innovation, School of Engineering, The University of Tokyo, Tokyo, Japan
e-mail: fujita@ipr-ctr.t.u-tokyo.ac.jp

Takayuki Ito
School of Techno-Business Administration, Nagoya Institute of Technology, Nagoya, Aichi, Japan
e-mail: ito.takayuki@nitech.ac.jp

Mark Klein
Sloan School of Management, Massachusetts Institute of Technology, Cambridge, MA, U.S.
e-mail: m_klein@mit.edu

T. Ito et al. (Eds.): Complex Automated Negotiations, SCI 435, pp. 23–39.
springerlink.com

1 Introduction

Negotiation is an important aspect of daily life and represents an important topic in the field of multi-agent system research. There has been extensive work in the area of automated negotiation; that is, where automated agents negotiate with other agents in such contexts as e-commerce[13], large-scale deliberation[20], collaborative design, and so on. Many real-world negotiations are complex and involve interdependent issues. When designers work together to design a car, for example, the utility of a given carburetor choice is highly dependent on which engine is chosen. The key impact of such issue dependencies is that they create qualitatively more complex utility functions, with multiple optima. There has been an increasing interest in negotiation with multiple interdependent issues. [9, 17, 21, 22, 24]. To date, however, achieving high scalability in negotiations with multiple interdependent issues remains an open problem.

We propose a new protocol in which a mediator tries to reorganize a highly complex utility space with issue interdependencies into several tractable subspaces, in order to reduce the computational cost. We call these utility subspaces "Issue groups." First, the agents generate interdependency graphs which capture the relationships between the issues in their individual utility functions, and derive issue clusters from that. While others have discussed issue interdependency in utility theory[26, 2], these efforts weren't aimed at efficiently decomposing the contract space. Second, the mediator combines these issue clusters to identify aggregate issue groups. Finally, the mediator uses a nonlinear optimization protocol to find sub-agreements for each issue group based on votes from the agents, and combines them to produce the final agreement.

We also address a negotiation between Exaggerator Agents. Any voting scheme introduces the potential for strategic non-truthful voting by the agents, and our method is no exception. For example, one of the agents may always vote truthfully, while the other exaggerates so that its votes are always "strong." It has been shown that this biases the negotiation outcomes to favor the exaggerator, at the cost of reduced social welfare. We employ the limitation of strong votes to the issue-grouping method. We investigate whether this approach can be applied to the method of decomposing a contract space.

The remainder of this paper is organized as follows. We describe a model of multiple interdependent issues negotiation and the strength of interdependency between issues, and the structure of interdependency graph. Next, we present a clustering technique for finding issue sub-groups. We then propose a protocol that uses this issue group information to enable more scalable negotiations. We also describe the effect of Exaggerator Agents in multi-agent situations. We present the experimental results, demonstrating that our protocol produces more optimal outcomes than previous efforts. Finally, we describe related work and present our overall conclusions.

2 Negotiation with Nonlinear Utility Functions

2.1 Multi-issue Negotiation Model

We consider the situation where N agents $(a_1,\ldots,a_N)$ want to reach an agreement with a mediator who manages the negotiation from a man-in-the-middle position. There are M issues $(i_1,\ldots,i_M)$ to be negotiated. The number of issues represents the number of dimensions in the utility space. The issues are shared: all agents are potentially interested in the values for all M issues. A contract is represented by a vector of values $\mathbf{s} = (s_1,...,s_M)$. Each issue s_j has a value drawn from the domain of integers $[0,X]$, *i.e.*, $s_j \in \{0,1,,\ldots,X\}(1 \leq j \leq M)$. [1].

An agent's utility function, in our formulation, is described in terms of constraints. There are l constraints, $c_k \in C$. Each constraint represents a volume in the contract space with one or more dimensions and an associated utility value. c_k has value $w_a(c_k,\mathbf{s})$ if and only if it is satisfied by contract $\mathbf{s}$. Function $\delta_a(c_k,i_j)$ is a region of i_j in c_k, and $\delta_a(c_k,i_j)$ is $\emptyset$ if c_k doesn't have any relationship to i_j. Every agent has its own, typically unique, set of constraints.

An agent's utility for contract $\mathbf{s}$ is defined as the sum of the utility for all the constraints the contract satisfies, i.e., as $u_a(\mathbf{s}) = \sum_{c_k \in C,\mathbf{s} \in x(c_k)} w_a(c_k,\mathbf{s})$, where $x(c_k)$ is a set of possible contracts (solutions) of c_k. This formulation produces complex utility functions with high points where many constraints are satisfied and lower regions where few or no constraints are satisfied. Many real-world utility functions are quite complex in this way, involving many issues as well as higher-order (e.g. trinary and quaternary) constraints. This represents a crucial departure from most previous efforts on multi-issue negotiation, where contract utility has been calculated as the weighted sum of the utilities for individual issues, producing utility functions shaped like hyper-planes, with a single optimum.

Figure 1 shows an example of a utility space generated via a collection of binary constraints involving Issues 1 and 2. In addition, the number of terms is two. The example, which has a value of 55, holds if the value for Issue 1 is in the range $[3,7]$ and the value for Issue 2 is in the range $[4,6]$. The utility function is highly nonlinear with many hills and valleys. This constraint-based utility function representation allows us to capture the issue interdependencies common in real-world negotiations. The constraint in Figure 1, for example, captures the fact that a value of 4 is desirable for issue 1 if issue 2 has the value 4, 5 or 6. Note, however, that this representation is also capable of capturing linear utility functions as a special case (they can be captured as a series of unary constraints). A negotiation protocol for complex contracts can, therefore, handle linear contract negotiations.

This formulation was described in [9]. In [17, 21, 22], a similar formulation is presented that supports a wider range of constraint types.

[1] A discrete domain can come arbitrarily close to a 'real' domain by increasing its size. As a practical matter, many real-world issues that are theoretically 'real' numbers (delivery date, cost) are discretized during negotiations.

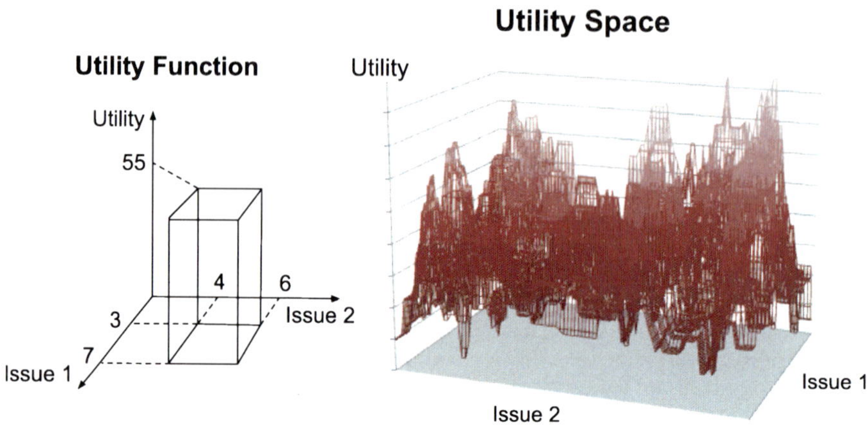

Fig. 1 Example of a nonlinear utility space

The objective function for our protocol can be described as follows:

$$\arg\max_{\mathbf{s}} \sum_{a \in N} u_a(\mathbf{s}). \tag{1}$$

$$\arg\max_{\mathbf{s}} u_a(\mathbf{s}),\ (a = 1, \ldots, N). \tag{2}$$

Our protocol, in other words, tries to find contracts that maximize social welfare, *i.e.*, the summed utilities for all agents. Such contracts, by definition, will also be Pareto-optimal. At the same time, all the agent try to find contracts that maximize their own welfare.

3 Our Negotiation Protocol: Decomposing the Contract Space

It is of course theoretically possible to gather all of the individual agents' utility functions in one central place and then find all optimal contracts using such well-known nonlinear optimization techniques as simulated annealing or evolutionary algorithms. However, we do not employ such centralized methods for negotiation purposes because we assume, as is common in negotiation contexts, that agents prefer not to share their utility functions with each other, in order to preserve a competitive edge.

Our approach is described in the following sections.

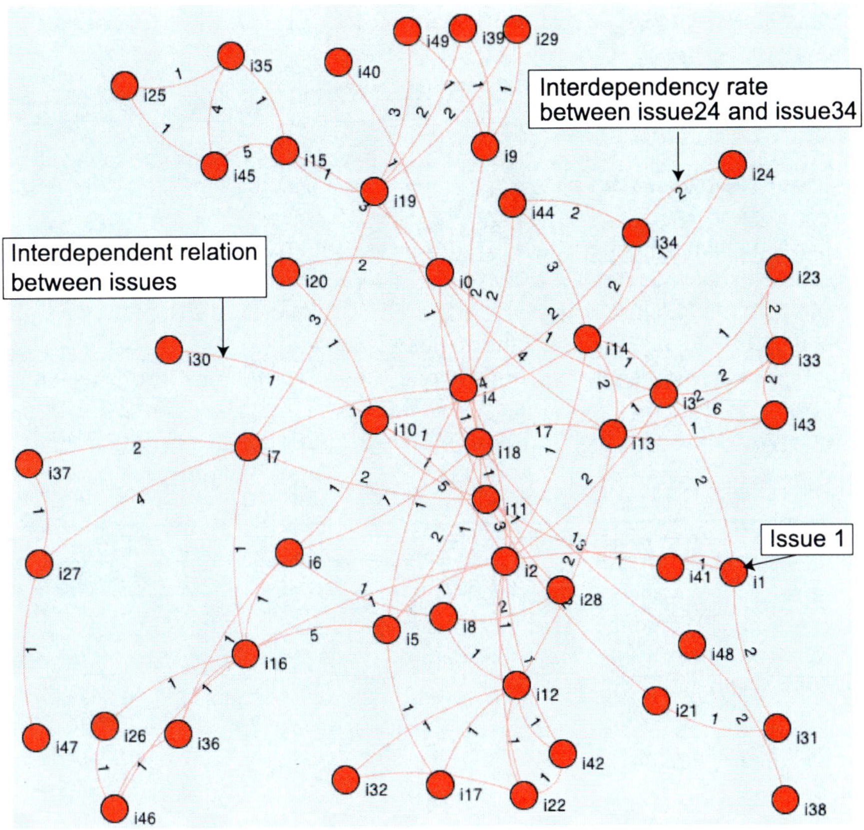

Fig. 2 Interdependency Graph (50 issues)

3.1 Analyzing Issue Interdependency

The first step is for each agent to generate an interdependency graph by analyzing the issue interdependencies in its own utility space. We define issue interdependency as follows. If there is a constraint between issue X (i_X) and issue Y (i_Y), then we assume i_X and i_Y are interdependent. If, for example, an agent has a binary constraint between issue 1 and issue 3, those issues are interdependent for that agent.

The *strength* of an issue interdependency is captured by the interdependency rate. We define the interdependency rate between two issues as the number of constraints that inter-relate them. The interdependency rate between *issue* i_j and *issue* i_{jj} for agent a is thus $D_a(i_j, i_{jj}) = \sharp\{c_k | \delta_a(c_k, i_j) \neq \emptyset \wedge \delta_a(c_k, i_{jj}) \neq \emptyset\}$.

Agents capture their issue interdependency information in the form of interdependency graphs i.e. weighted non-directed graphs where a node represents an issue, an edge represents the interdependency between issues, and the weight of an edge represents the interdependency rate between those issues. An interdependency graph is thus formally defined as: $G(P,E,w) : P = \{1,2,\ldots,|I|\}(finite\ set), E \subset \{\{x,y\}|x,y \in P\}, w : E \rightarrow R$.

Figure 2 shows an example of an interdependency graph.

3.2 Grouping Issues

In this step, the mediator employs breadth-first search to combine the issue clusters submitted by each agent into a consolidated set of issue groups. For example, if agent 1 submits the clusters $\{i_1,i_2\},\{i_3,i_4,i_5\}$, $\{i_0,i_6\}$ and agent 2 submits the clusters $\{i_1,i_2,i_6\},\{i_3,i_4\},\{i_0\},\{i_5\}$, the mediator combines them to produce the issue groups $\{i_0,i_1,i_2,i_6\}$, $\{i_3,i_4,i_5\}$. In the worst case, if all the issue clusters submitted by the agents have overlapping issues, the mediator generates the union of the clusters from all the agents. The details of this algorithm are given in Algorithm 1.

Listing 1. Combine_IssueGroups(G)

Ag: A set of agents, G: A set of issue-groups of each agent ($G = \{G_0, G_1, ..., G_n\}$, a set of issue-groups from agent i is $G_i = \{g_{i,0}, g_{i,1}, ..., g_{i,m_i}\}$)

```
SG := G_0, i := 1
while i < |Ag| do
    SG' := ∅
    for s ∈ SG do
        for g_{i,j} ∈ G_i do
            s' := s ∩ g_{i,j}
            if s' ≠ φ then
                SG' := s ∪ g_{i,j}
            end if
            SG := SG', i := i + 1
        end for
    end for
end while
```

It is possible to gather all of the agents' interdependency graphs in one central place and then find the issue groups using standard clustering techniques. However, it is hard to determine the optimal number of issue groups or the clustering parameters in central clustering algorithms, because the basis of clustering for every agent can be different. Our approach avoids these weaknesses by requiring that each agent generates its own issue clusters. In our experiments, agents did so using the well-known Girvan-Newman algorithm[18], which computes clusters in weighted non-direct graphs. The algorithm's output can be controlled by changing the "number of edges to remove" parameter. Increasing the value of this parameter increases the number of issue dependencies ignored when calculating the issue clusters, thereby resulting in a larger number of smaller clusters. The running time of this algorithm is

$O(kmn)$, where k is the number of edges to remove, m is the total number of edges, and n is the total number of vertices.

3.3 *Finding Agreements*

We use a distributed variant of simulated annealing (SA)[11] to find optimal contracts in each issue group. In each round, the mediator proposes a contract that is a random single-issue mutation of the most recently accepted contract (the accepted contract is initially generated randomly). Each agent then votes to accept(+2), weakly accept(+1), weakly reject(-1) or reject(-2) the new contract, based on whether it is better or worse than the last accepted contract for that issue group. When the mediator receives these votes, it adds them together. If the sum of the vote values from the agents is positive or zero, the proposed contract becomes the currently accepted one for that issue group. If the vote sum is negative, the mediator will accept the contract with probability $P(accept) = e^{\Delta U/T}$, where T is the mediator's virtual temperature (which declines over time) and ΔU is the utility change between the contracts. In other words, the higher the virtual temperature, and the smaller the utility decrement, the greater the probability that the inferior contract will be accepted. If the proposed contract is not accepted, a mutation of the most recently accepted contract is proposed in the next round. This continues over many rounds. This technique allows the mediator to skip past local optima in the utility functions, especially earlier on in the search process, in the pursuit of global optima.

Listing 2. Simulated_Annealing()

Value(N): the sum of the numeric values mapped from votes to N from all agents

```
S := initial solution (set randomly)
for t = 1 to ∞ do
   T := schedule(t)
   if T = 0 then
      return current
   end if
   next := a randomly selected successor of current
   if next.Value ≥ 0 then
      ΔE := next.Value − current.Value
      if ΔE > 0 then
         current := next
      else
         current := next only with probability e^(ΔE/T)
      end if
   end if
end for
```

3.4 Exaggerator Agents

Any voting scheme introduces the potential for strategic non-truthful voting by the agents, and our method is no exception. For example, one of the agents may always vote truthfully, while the other exaggerates so that its votes are always "strong." It has been shown that this biases the negotiation outcomes to favor the exaggerator, at the cost of reduced social welfare [12]. What we need is an enhancement of our negotiation protocol that preventing the exaggerator votes and maximizing social welfare.

We guess that simply placing a limit on the number of "strong" votes each agent can work well. If the limit is too low, we effectively lose the benefit of vote weight information and get the lower social welfare values that result. If the strong vote limit is high enough to avoid this, then all an exaggerator has to do is save all of its strong votes until the end of the negotiation, at which point it can drag the mediator towards making a series of proposals that are inequitably favorable to it. In the experiments, we demonstrate that the limit of the number of "strong" voting is efficient of finding high solutions.

4 Experimental Results

4.1 Setting

We conducted several experiments to evaluate our approach. In each experiment, we ran 100 negotiations. The following parameters were used. The domain for the issue values was $[0,9]$. Each agent had 10 unary constraints, 5 binary constraints, 5 trinary constraints, and so on. (a unary constraint relates to one issue, a binary constraint relates to two issues, etc). The maximum weight for a constraint was *100* $\times$ *(Number of Issues)*.

In our experiments, each agents' issues were organized into ten small clusters with strong dependencies between the issues within each cluster. We ran two conditions: "1) Sparse Connection" and "2) Dense Connection". Figure 3 gives examples, for these two cases, of interdependency graphs and the relationship between the number of issues and the sum of the connection weights between issues. As these graphs show, the "1) Sparse Connection" case is closer to a scale-free distribution, with power-law statistics, while the "2) Dense connection" is closer to a random graph.

We compared the following negotiation methods:

"(A) Issue-Grouping (True Voting)" applies the simulated annealing protocol based on the agents' votes, and performs the negotiation separately for each one of the issue groups, and combines the resulting sub-agreements to produce the final agreement. All agents tell the truth votes. "(B) Issue-Grouping (Exaggerator Agents)" applies the simulated annealing protocol based on the agents' votes with

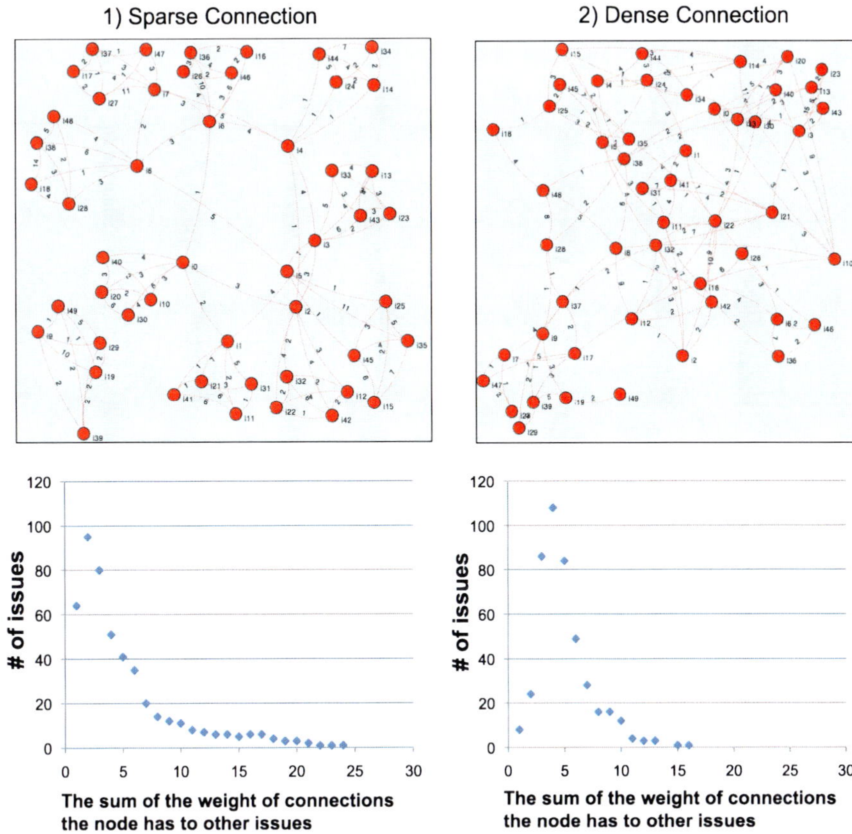

Fig. 3 Issue Interdependencies

issue-grouping. "All agent" tell the exaggerator votes. "(C) Issue-Grouping (limitation)" is same situation with (B). However, the limitation of 'strong' votes is applied. The number of limitation of 'strong' votes is 250 which is the optimal number of limitations in this experiments. "(D) Without Issue-Grouping" is the method presented in Klein et.al[12], using a simulated annealing protocol based on the agents' votes without generating issue-groups.

In all these cases, the search began with a randomly generated contract, and the SA initial temperature for all these cases was 50.0 and decreased linearly to 0 over the course the negotiation. In case (D), the search process involved 500 iterations. In case (A)-(C), the search process involved 50 iterations for each issue group. Cases (A),(B),(C) and (D) thus used the same amount of computation time, and are thus directly comparable. The number of edges removed from the issue interdependency graph, when the agents were calculating their issue groups, was 6 in all cases.

We applied a centralized simulated annealing to the sum of the individual agents' utility functions to approximate the optimal social welfare for each negotiation test

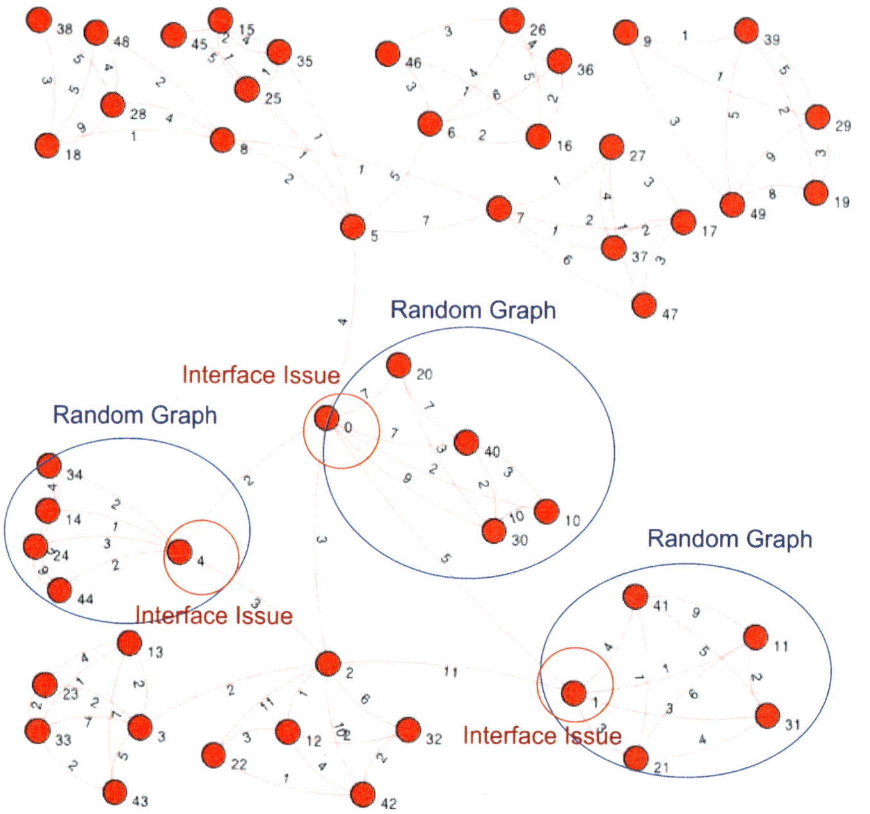

Fig. 4 Method of determining interdependency graph

run. Exhaustive search was not a viable option because it becomes computationally intractable as the number of issues grows. The SA initial temperature was 50.0 and decreases linearly to 0 over the course of 2,500 iterations. The initial contract for each SA run is randomly selected. We calculated a normalized "optimality rate" for each negotiation run, defined as *(social welfare achieved by each protocol) / (optimal social welfare calculated by SA)*.

Our code was implemented in Java 2 (1.6) and was run on a core 2-duo CPU with 2.0 GB memory under Mac OS X (10.6).

4.2 Method of Determining Interdependency Graph

Figure 4 shows what the interdependency graph consists of in an agent.

The method of determining the interdependency between issues in the experiment is as follows.

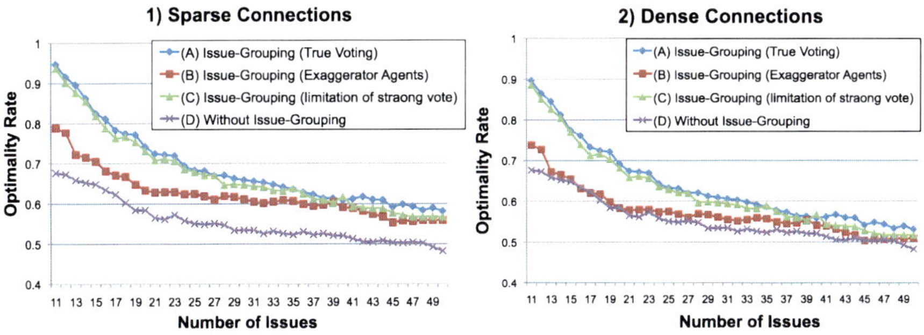

Fig. 5 Comparison of optimality when the number of issues changes

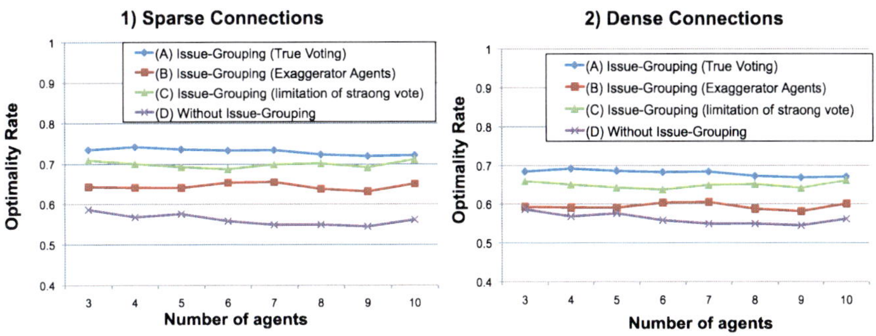

Fig. 6 Comparison of optimality when the number of agents changes

(Step 1) Small issue-groups are generated by connecting a part of the issues randomly.

(Step 2) The interface issues are decided randomly among issues in each issue-group. The interface issues are for connecting other small issue-groups. In small issue-groups, only the interface issues can connect to other issue-groups.

(Step 3) Each issue-group connects to other small issue-groups. Specifically, all combinations of each issue-group are searched for, and it is decided whether connection or disconnection according to the possibility of generating connections.

4.3 Experimental Results

Figure 5 and 6 compare the optimality rate in the sparse connection and dense connection cases. "(A) Issue-Grouping (True Voting)" achieved a higher optimality rate than "(D) Without Issue-Grouping" which means that the issue-grouping method produces better results for the same amount of computational effort. The optimality rate of the "(A) Issue-Grouping (True Voting)" condition decreased as the number of issues (and therefore the size of the search space) increased. "(B) Issue-Grouping (Exaggerator Agents)" is worse than "(A) Issue-Grouping (True Voting)" because

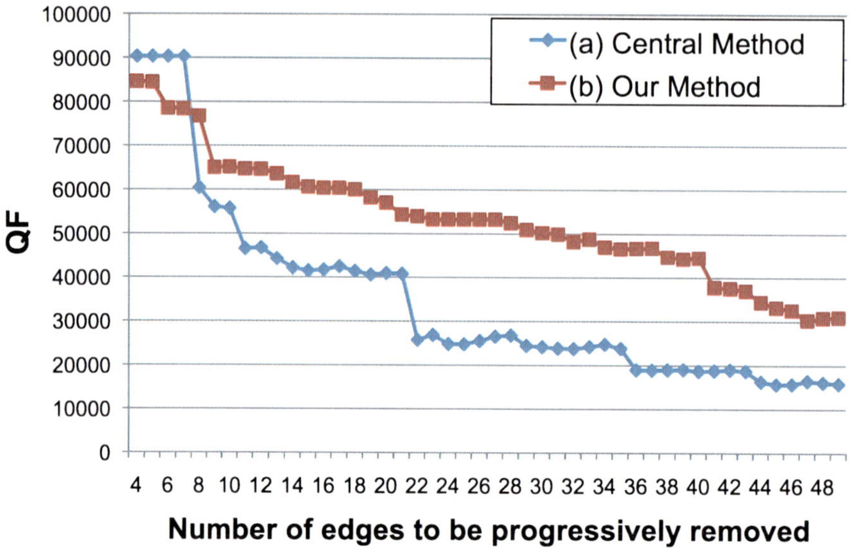

Fig. 7 Number of edges to be progressively removed (Clustering parameter) v.s. QF

the exaggerator agents generate reduced social welfares in multi-agents situations. However, "(C) Issue-Grouping (limitation)" outperforms "(B) Issue-Grouping (Exaggerator Agents)", therefore, the limitation of 'strong' votes is effective of improving the social welfare reduced by the Exaggerator Agents.

The optimality rates for all methods are almost unaffected by the number of agents, as Figure 6 shows. The optimality rate for (A) is higher than (D) in the "1) Sparse Connections" case than the "2) Dense Connections" case. This is because the issue grouping method proposed in this paper can achieve high optimality if the number of ignored interdependencies is low, which is more likely to be true in the "1) Sparse Connections" case. Many real-world negotiations are, we believe, characterized by sparse issue inter-dependencies.

We also assessed a quality factor measure *QF = (Sum of internal weights of edges in each issue-group) / (Sum of external weights of edges in each issue-group)* to assess the quality of the issue groups, i.e. the extent to which issue dependencies occurred only between issues in the same clusters, rather than between issues in different groups. Higher quality factors should, we predict, increase the advantage of the issue grouping protocols, because that means fewer dependencies are ignored when negotiation is done separately for each issue group. Figure 7 shows the quality factors when the number of agents is 3 and 20, as a function of the number of edges to be removed (which is the key parameter in the clustering algorithm we used).The number of issues is 50 in the "1) sparse connection" case. "(a) Central Method" is to gather all of the agents' interdependency graphs in one central place and then find the issue groups using the well-known Girvan-Newman algorithm[18]. "(b)

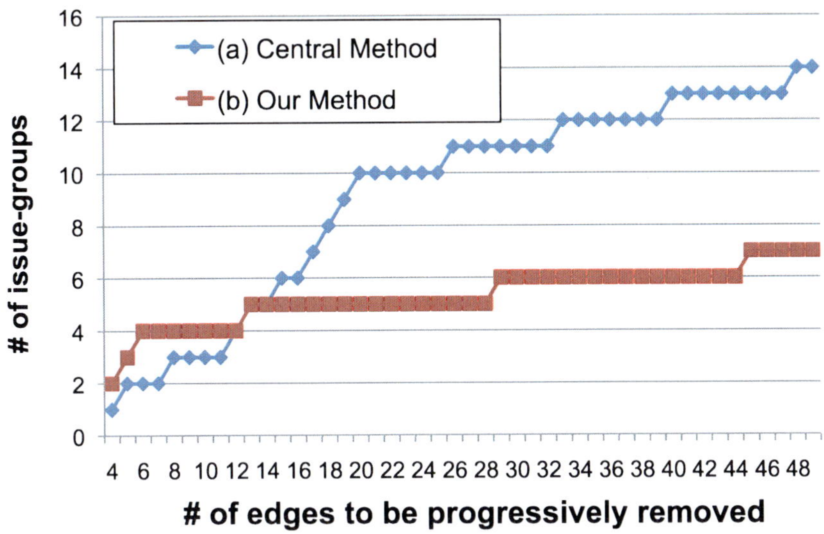

Fig. 8 Number of edges to be progressively removed (Clustering parameter) v.s. The number of issue-groups

Our method" employs breadth-first search to combine the issue clusters submitted by each agent into a consolidated set of issue groups.

Comparing (a) with (b) in Figure 7, (b) proposed in this paper outperforms (a). This is because that our method is reflected by the idea of all agents to final issue-grouping without fixing the clustering parameter as Figure8 showing. QF becomes smaller when the number of edges to be progressively removed is larger. This is because the number of issue-groups generated by each agent is higher as the number of edges to be progressively removed becomes larger. The rapid decrease sometimes happens as the number of edges to be progressively removed increases. These points are good parameters for decomposing the issue-groups. In real life, the utility of agents contains an adequate idea of issue-groups, and agents can determine the optimal idea of issue-groups by analyzing the utility spaces.

5 Related Work

Even though negotiation seems to involve a straightforward distributed constraint optimization problem [7, 19], we have been unable to exploit existing work on high-efficiency constraint optimizers. Such solvers attempt to find the solutions that maximize the weights of the satisfied constraints, but do not account for the fact that the final solution must satisfy at least one constraint *from every agent*.

Lin et al.[16] explored a range of protocols based on mutation and selection on binary contracts. This paper does not describe what kind of utility function is used, nor does it present any experimental analyses, so it remains unclear whether this strategy enables sufficient exploration of utility space.

Klein et al.[12] presented a protocol applied with near optimal results to medium-sized bilateral negotiations with binary dependencies, but was not applied to multilateral negotiations and higher order dependencies.

A bidding-based protocol was proposed by Ito et al.[9]. Agents generate bids by finding high regions in their own utility functions, and the mediator finds the optimum combination of submitted bids from the agents. However, the scalability of this protocol is limited, and the failure rate of making agreements is too high. By Fujita et al.[5], a representative-based protocol for reducing the computational cost was proposed based on the bidding-based protocol. In this method, the scalability of agents was improved; however, the scalability of issues was not sufficient. Fujita et.al[6] also focused on the decomposing the contract space for highly scalable negotiation, but the negotiation protocol and experimental results are completely different.

Hindriks et al.[8] proposed an approach based on a weighted approximation technique to simplify the utility space. The resulting approximated utility function without dependencies can be handled by negotiation algorithms that can efficiently deal with independent multiple issues, and has a polynomial time complexity. Our protocol can find an optimal agreement point if agents don't have in common the expected negotiation outcome.

Fatima et al.[3, 4] proposed bilateral multi-issue negotiations with time constraints. This method can find approximate equilibrium in polynomial time where the utility function is nonlinear. However, this paper focused on bilateral multi-issue negotiations. Our protocol focuses on multilateral negotiations.

Zhang[27] presents an axiomatic analysis of negotiation problems within task-oriented domains (TOD). In this paper, three classical bargaining solutions (Nash solution, Egalitarian solution, Kalai-Smorodinsky solution) coincide when they are applied to a TOD with mixed deals but diverge if their outcomes are restricted to pure deals.

Maestre et al.[21, 22, 23] proposed an auction-based protocol for nonlinear utility spaces generated using weighted constraints, and proposed a set of decision mechanisms for the bidding and deal identification steps of the protocol. They proposed the use of a quality factor to balance utility and deal probability in the negotiation process. This quality factor is used to bias bid generation and deal identification, taking into account the agents' attitudes toward risk. The scalability of the number of issues is still a problem in these works.

Jonker et al.[10] proposed a negotiation model called ABMP that can be characterized as cooperative one-to-one multi-criteria negotiation in which the privacy of both parties is protected as much as desired.

By Robu et al.[24], utility graphs were used to model issue dependencies for binary-valued issues. Our utility model is more general.

Bo et al.[1] proposed the design and implementation of a negotiation mechanism for dynamic resource allocation problem in cloud computing. Multiple buyers and sellers are allowed to negotiate with each other concurrently and an agent is allowed to decommitment from an agreement at the cost of paying a penalty.

Lin et al. [14, 15] focus on the Expert Designed Negotiators (EDN) which is the negotiations between humans and automated agents in real-life. In addition, the tools for evaluating automatic agents that negotiate with people were proposed. These studies include some efficient results from extensive experiments involving many human subjects and PDAs.

6 Conclusion

In this paper, we proposed a new negotiation protocol, based on grouping issues, which can find high-quality agreements in interdependent issue negotiation. In this protocol, agents generate their private issue interdependency graphs and use these to generate issue clusters. The mediator consolidates these clusters to define aggregate issue groups, and independent negotiations proceed for each group. We analyzed the negotiation that one of agents may always vote truthfully, while the other exaggerates so that its votes are always "strong." We demonstrated that our proposed protocol results in a higher optimality rate than methods that don't use issue grouping, especially when the issue interdependencies are relatively sparse. In addition, the limitation of "strong" votes is effective of improving the reduced social welfare in multi-agent negotiations between exaggerators.

In future work, we will conduct additional negotiation, after the concurrent subcontract negotiations, to try to increase the satisfaction of constraints that crossed issue group boundaries and were thus ignored in our issue grouping approach. In the bilateral case, we found this can be done using a kind of Clarke tax [25], wherein each agent has a limited budget from which it has to pay other agents before the mediator will accept a contract that favors that agent but reduces utility for the others. We investigate whether and how this approach can be applied to the multilateral case.

References

1. An, B., Lesser, V.R., Irwin, D., Zink, M.: Automated negotiation with decommitment for dynamic resource allocation in cloud computing. In: Proc. of the Ninth International Joint Conference on Autonomous Agents and Multi-agent Systems, AAMAS 2010, pp. 981–988 (2010)
2. Eppinger, S.D.: Model-based approaches to managing concurrent engineering. Journal of Engineering Design 2(4), 283–290 (1991)
3. Fatima, S.S., Wooldridge, M., Jennings, N.R.: An analysis of feasible solutions for multi-issue negotiation involving nonlinear utility functions. In: Proc. of the Eighth International Joint Conference on Autonomous Agents and Multi-agent Systems, AAMAS 2009, pp. 1041–1048 (2007)

4. Fatima, S.S., Wooldridge, M., Jennings, N.R.: Approximate and online multi-issue negotiation. In: Proc. of the Sixth International Joint Conference on Autonomous Agents and Multi-agent Systems, AAMAS 2007, pp. 947–954 (2007)
5. Fujita, K., Ito, T., Klein, M.: A representative-based multi-round protocol for multi-issue negotiations. In: Proc. of the Seventh Inernational Joint Conference on Autonomous Agents and Multi-agent Systems, AAMAS 2008, pp. 1573–1576 (2008)
6. Fujita, K., Ito, T., Klein, M.: An approach to scalable multi-issue negotiation: Decomposing the contract space based on issue interdependencies. In: Proc. of the 2010 International Joint Conference on Intelligent Agent Technology, IAT 2010 (2010)
7. Greenstadt, R., Pearce, J., Tambe, M.: Analysis of privacy loss in distributed constraint optimization. In: Proc. of the 21st Association for the Advancement of Artificial Intelligence, AAAI 2006, pp. 647–653 (2006)
8. Hindriks, K.V., Jonker, C.M., Tykhonov, D.: Eliminating Interdependencies Between Issues for Multi-issue Negotiation. In: Klusch, M., Rovatsos, M., Payne, T.R. (eds.) CIA 2006. LNCS (LNAI), vol. 4149, pp. 301–316. Springer, Heidelberg (2006)
9. Ito, T., Hattori, H., Klein, M.: Multi-issue negotiation protocol for agents: Exploring nonlinear utility spaces. In: Proc. of the 20th International Joint Conference on Artificial Intelligence, IJCAI 2007, pp. 1347–1352 (2007)
10. Jonker, C.M., Robu, V., Treur, J.: An agent architecture for multi-attribute negotiation using incomplete preference information. Journal of Autonomous Agents and Multi-Agent Systems (JAAMAS) 15, 221–252 (2007)
11. Kirkpatrick, S., Gelatt, C.D., Vecchi, M.P.: Optimization by Simulated Annealing. Science 220(4598), 671–680 (1983)
12. Klein, M., Faratin, P., Sayama, H., Bar-Yam, Y.: Negotiating complex contracts. Group Decision and Negotiation 12(2), 58–73 (2003)
13. Kraus, S.: Strategic Negotiation in Multiagent Environments. Cambridge University Press (2001)
14. Lin, R., Kraus, S.: Can automated agents proficiently negotiate with humans? Commun. ACM 53(1), 78–88 (2010)
15. Lin, R., Kraus, S., Oshrat, Y., Gal, Y.K.: Facilitating the evaluation of automated negotiators using peer designed agents. In: Proc. of The 24th Association for the Advancement of Artificial Intelligence, AAAI 2010 (2010)
16. Lin, R.J., Chou, S.T.: Bilateral multi-issue negotiations in a dynamic environment. In: Proc. of the AAMAS Workshop on Agent Mediated Electronic Commerce, AMEC 2003 (2003)
17. Lopez-Carmona, M., Marsa-Maestre, I., Klein, M., Ito, T.: Addressing stability issues in mediated complex contract negotiations for constraint-based, non-monotonic utility spaces. In: Autonomous Agents and Multi-Agent Systems, pp. 1–51 (2010)
18. Girvan, M., Newman, M.E.J.: Community structure in social and biological networks. National Academy of Sciences of USA (PNAS) 99(12), 7821–7826 (2002)
19. Maheswaran, R.T., Pearce, J.P., Varakantham, P., Bowring, E.: Valuations of possible states (vps):a quantitative framework for analysis of privacy loss among collaborative personal assistant agents. In: Proc. of the Fourth International Joint Conference on Autonomous Agents and Multi-agent Systems, AAMAS 2005, pp. 1030–1037 (2005)
20. Malone, T.W., Klein, M.: Harnessing collective intelligence to address global climate change. Innovations Journal 2(3), 15–26 (2007)
21. Marsa-Maestre, I., Lopez-Carmona, M.A., Velasco, J.R., de la Hoz, E.: Effective bidding and deal identification for negotiations in highly nonlinear scenarios. In: Proc. of the Eighth International Joint Conference on Autonomous Agents and Multi-Agent Systems, AAMAS 2009, pp. 1057–1064 (2009)
22. Marsá-Maestre, I., López-Carmona, M.A., Velasco, J.R., de la Hoz, E.: Avoiding the prisoner's dilemma in auction-based negotiations for highly rugged utility spaces. In: Proc. of the Ninth International Joint Conference on Autonomous Agents and Multi-agent Systems, AAMAS 2010, pp. 425–432 (2010)

23. Marsa-Maestre, I., Lopez-Carmona, M.A., Velasco, J.R., Ito, T., Klein, M., Fujita, K.: Balancing utility and deal probability for negotiations in highly nonlinear utility spaces. In: Proc. of the Twenty-first International Joint Conference on Artificial Intelligence, IJCAI 2009, pp. 214–219 (2009)
24. Robu, V., Somefun, D.J.A., Poutre, J.L.: Modeling complex multi-issue negotiations using utility graphs. In: Proc. of the Forth International Joint Conference on Autonomous Agents and Multi-Agent Systems, AAMAS 2005, pp. 280–287 (2005)
25. Sandholm, T.W.: Distributed Rational Decision Making. In: Weiss, G. (ed.) Multi-Agent Systems (1998)
26. Tamura, H., Nakamura, Y.: Decompositions of multiattribute utility functions based on convex dependence. Operations Research 31(3), 488–506 (1983)
27. Zhang, D.: Axiomatic characterization of task oriented negotiation. In: Proc. of Twenty-first International Joint Conference on Artificial Intelligence, IJCAI 2009, pp. 367–372 (2009)

Efficient Deal Identification for the Constraints Based Utility Space Model

Raiye Hailu and Takayuki Ito

Abstract. We propose correct and efficient algorithms for locating the optimal contract of negotiating agents that represent their utility space with the constraints based utility space model. It is argued that the agents that use the model can be classified in to two extreme kinds: sensitive and insensitive. When the negotiation is between a sensitive agent and many in- sensitive agents, the optimal contract can be computed correctly and efficiently by avoiding Exhaustive Matching.

1 Introduction

Automating negotiations over multiple and interdependent issues is potentially an important line of research since most negotiations in the real world have interdependent issues. When a service provider negotiates on When to provide its service, its utility for a certain time period (e.g. T1=8a.m- 10a.m) is dependent on the day of the week (Monday-Sunday). It might have high utility for T1 on Mondays, but low utility for T1 on Sundays. The issues, time of the meeting and day of the meeting cannot be negotiated independently.

We propose correct and efficient algorithm for locating the optimal contract of negotiating agents that represent their utility space with the constraints based utility space model proposed in [4]. The model is used to represent utility space of agents negotiating over multiple and interdependent issues. Some researchers [1, 2, 3, 5] have proposed algorithms(protocols) for locating the optimal contract. The proposed

Raiye Hailu
Department of Computer Science and Engineering, Nagoya Institute of Technology
e-mail: raiye@itolab.nitech.ac.jp

Takayuki Ito
School of Techno-Business Administration, Nagoya Institute of Technology
e-mail: ito.takayuki@nitech.ac.jp

T. Ito et al. (Eds.): Complex Automated Negotiations, SCI 435, pp. 41–53.
springerlink.com

algorithms have their own merits, but they all fall under the classification of heuristic algorithms when evaluated solely from the view point of locating the optimal contract correctly. The optimal contract is the contract that has the maximum total utility. Total utility for a contract is the sum of the utility of each agent for the contract.

Exhaustively Matching (EM) the entire utility space of the agents is the only correct method of searching the optimal contract. If the utility space of agents is assumed to be generated randomly, then there is no method of making EM efficient (faster) and still guarantee correctness. Therefore we make intuitive assumptions about utility space of agents that can be readily implemented by the basic building block of the model - integer interval.

1.1 Constraints Based Utility Space Model

In the model, for agents negotiating on I number of issues, an I dimensional coordinate system is created. An axis is assigned to each issue. Each issue will have up to V number of issue values. We represent these values by integers ranging from 0 to V-1. Since the issues are interdependent, we will have V^I number of possible issue value combinations which are called contracts. An example of a contract is [0,2,4]. 0 is the issue value for I1(Issue 1) , 2 is the issue value for I2 etc.

The utility of a contract is the sum of the weights of all constraints satisfied at it. The constraint in Figure 1 has a weight of 55. Contracts that have the values 4 and 5 for issue 1, and the values 3, 4, 5 and 6 for issue 2 satisfy this constraint. An agent creates its utility space by defining multiple such constraints. Figure 2 shows a utility space created by using more than 100 constraints.

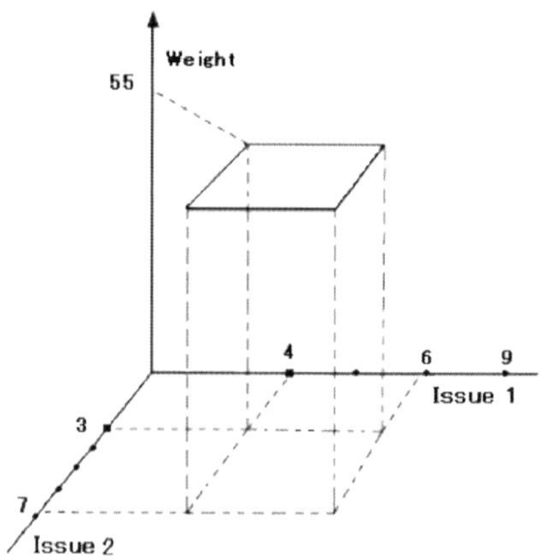

Fig. 1 A 2 issue constraint

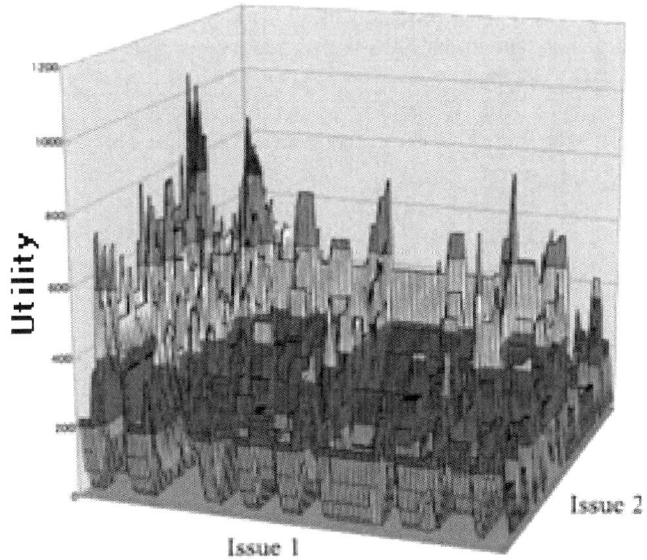

Fig. 2 A 2 issue utility space

2 Bidding Based Algorithm

Most previous works that used the constraints based utility space model use the bidding based deal identification method. Bids are high utility regions of the utility space of an agent. In a nutshell, bidding is the process of identifying and then submitting these high utility regions to a mediator agent. The mediator agent matches the bids to find those that have intersections and maximize the total utility. It was first proposed in [4]. Since then, some researchers have improved the method to address various concerns.

The threshold adjusting algorithm [1] makes agents bid in multiple rounds rather than once. In each round the threshold value is lowered. The threshold value is the minimum allowable utility value of a bid. The bidding is stopped at the round a deal is found. This has the advantage of limiting the amount of private information revealed to a third party.

The representative based algorithm [2] improves scalability of the bidding based algorithm by making only few agents called representatives participate in the bidding process. Scalability refers to the number agents that can be supported by the negotiation system. When the number of issues increases, the number of bids each agent has to make in order to effectively sample its utility space also increases. This in turn increases the time taken by the mediator to search an intersection of the bids that maximizes the total utility. If only the representatives are allowed to participate in the bid ding process, then negotiations with large number of agents can be supported.

When the contract space is large, the failure rate (when no bids from agents intersect) of a negotiation increases. The iterative narrowing protocol [3] reduces failure rates by narrowing down the region of the contract space that the agents generate their bids from. It is especially effective when the constraints of each agent are found being clustered in some of regions of the contract space, rather than being scattered all over the contract space.

Measures that reduce high failure rates that arise when agents use narrow constraints was discussed in [5]. The product of a bids utility and its volume was used as a criteria to select it to be submitted to the mediator or not. Usually high utility valued bids tend to be small in volume and therefore the chance that they will intersect with other agents' bids is minimal. Adding the volume criteria for selecting a bid for submission makes the deal identification process more effective. The problem is that the bid that contains the optimal contract may not be submitted by at least one of the agents. This might be because either that bid has low utility for that agent, or the bid generation mechanism "missed" it. Hence, there is always the chance that the optimal contract is not found.

3 Exhaustive Matching

The only way we can guarantee that the optimal contract is computed correctly is by making the agents submit their entire utility space to the mediator. Then the mediator Exhaustively Matches(EM) the utility spaces. The problem is that the computational time cost of this algorithm grows exponentially. If the number of issues of a negotiation grows from I to I + 1, then the contract space grows from V^I to V^{I+1}.

To reduce the time required to search for the optimal contract, we have to look for patterns in the utility space of agents that could be exploited to avoid EM. But observing Figure 1 and Figure 2 reveals that based on the number of constraints, their positioning and weight, utility spaces can be of various types and very unpredictable. The only predictable nature of them is that they are all based on constraints. Not just any constraint but integer interval based constraints.

3.1 Single Issue Version of the Model

The constraint in Figure 1 is a two dimensional integer interval of [4..5]x[3..6]. An example of a constraint in a negotiation over three issues would be [2..5]x[1..3]x[6..9]. If we were to define a single issue version of the model , then an example of a constraint would be [1..3]. Since the single issue version is easy to understand we will use it for analysis and experiments from here on wards. Since integer intervals are the basic building unit of the model we expect lessons learned from studying the single issue version of the model will be applicable for the multi issue version of it.

Figure 3 shows an agent that has 3 constraints :(C1, C2, C3). Its utility for the issue value 5 is: Weight (C2) + Weight (C3) = 10+20 = 30. Figure 4 is Figure 3 redrawn by summing the weights of each constraint. S0, S1...,S6 are called Steps of the utility function. Notice that Steps are also integer intervals. Also notice that, in a one issue utility space the issue values themselves are contracts of the negotiation. For example, in Figure 4, Step 4(S4) contains the contracts 4 and 5.

To avoid EM, we have to make assumptions about utility space of agents. To do that we still focus on integer intervals. This time the Steps are considered.

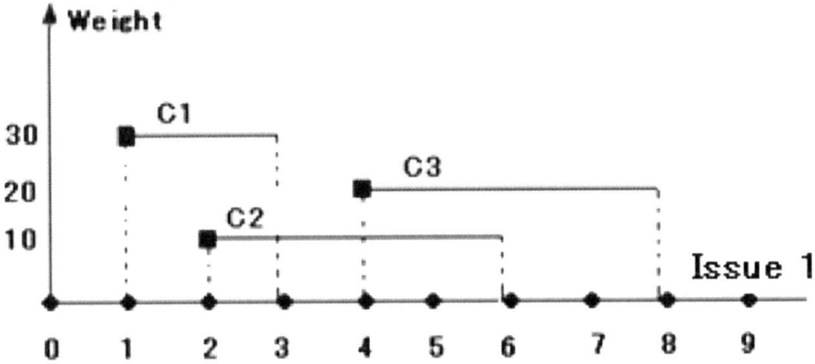

Fig. 3 Many single issue constraints

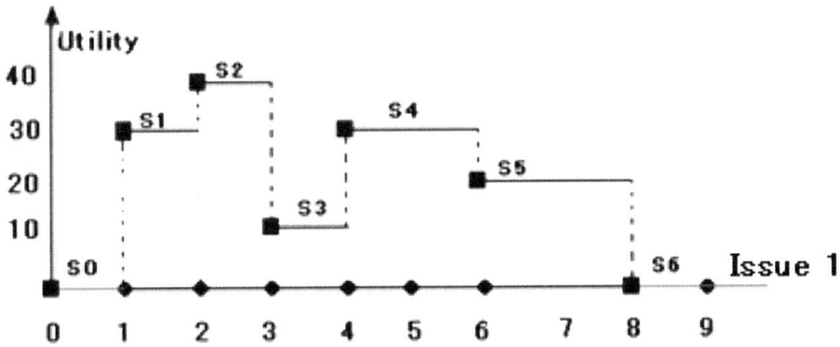

Fig. 4 Single issue utility space

3.2 *Sensitive and Insensitive Agents*

By focusing on the width of the Steps in the utility space of an agent, we can ask some interesting questions. If an agent's utility space is dominated by Steps that are wide, what does that say about the agent? What about when an agent's utility space is dominated by Steps that are narrow?

A Step contains consecutive contracts that the agent has equal utility for. Lets assume that consecutive contracts are more similar to one other than contracts that are far apart. Then, the fact that the agent has equal utility for some consecutive contracts indicates that, the agent neglects the small difference between the contracts. Based on this, we can classify agents to two extreme kinds: sensitive and insensitive. Here, the word, sensitive is used as it would be used for a sensor. A sensitive sensor is capable of registering small differences of the sensed signal.

Let's define a branch to be a portion of the contract space. For example, part of the contract space in Figure 4 containing the contracts 0 to 3 ([0..3]). In a branch, a sensitive agent will have four Steps. One for each contract. An insensitive agent will have one Step that contains all the contracts.(Currently we assume that the end points of the branches of all agents are the same and known).

Consider negotiation for scheduling a meeting of 30 minutes duration. A busy person is sensitive about every 30 minute interval. While he is relatively free at 10:30 a.m., he might have very important meeting at 11:00 a.m.. Therefore, he would not like to have the meeting at 11:00 a.m. (Figure 5). Hence, a busy persons utility space will be made of narrow width Steps. A free (not busy) person groups his time with large intervals (Figure 6). Hence, his utility space will be made up of wide Steps.

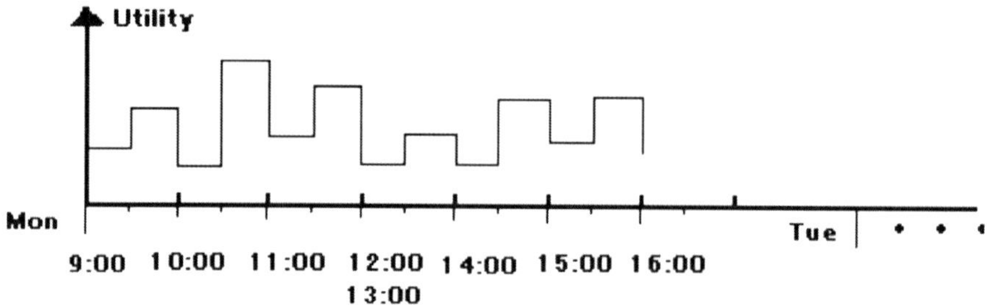

Fig. 5 Busy person

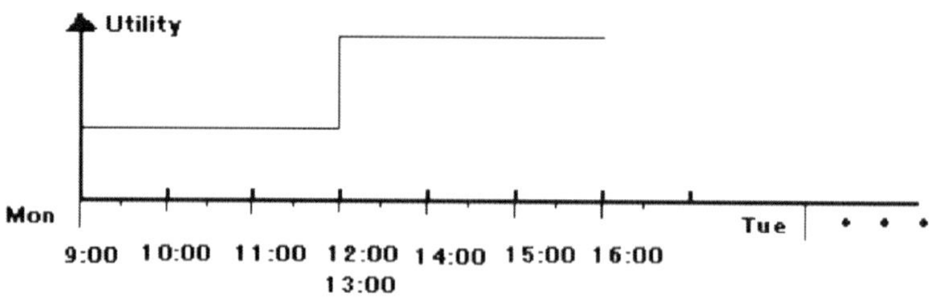

Fig. 6 A free(not busy) person

4 Exploiting Agents Difference in Their Sensitivity

4.1 COPE Algorithm

The COPE algorithm can locate the optimal contract more efficiently than EM when the COPE condition is satisfied. In Figure 7, a branch of a utility space is shown for four agents (Ag. h, i ,j and k). The optimal contract could be found by taking Step C of Ag. h (the Step with the highest utility) and matching it with the steps of Agents i,j and k. We call this method of computing the optimal contract COPE. Since the agents i,j and k have just one Step in the branch, just using the maximum Step of Ag. h is sufficient to correctly compute the optimal contract. For a branch the COPE condition is satisfied if,

1 Only The first agent in the matching lineup is sensitive; that is, it has many narrow width Steps.
2 The rest agents in the matching lineup have one wide Step which contains all the contracts in the branch.

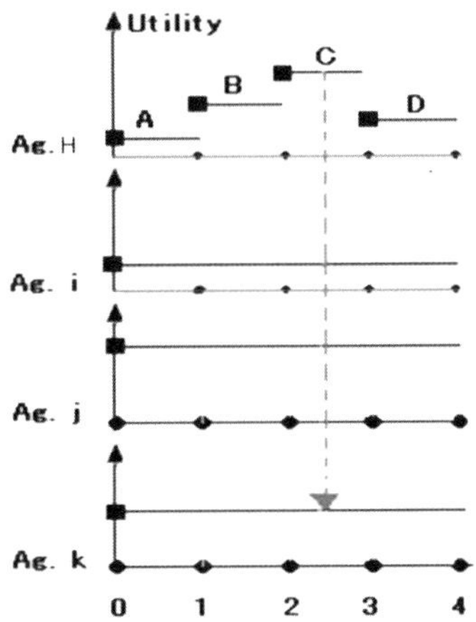

Fig. 7 COPE algorithm

4.2 FASTCOPE Algorithm

The COPE condition imposes stringent requirements on utility spaces of agents. One that could be relaxed is the requirement that the sensitive agent has to be the first in the matching line up. FASTCOPE algorithm is designed to compute the optimal contract efficiently even when the position of the sensitive agent is not known before hand. FASTCOPE algorithm extends COPE by rearranging the agents so that COPE condition is created before matching. The steps in the algorithm are:

Step 1 Identify the sensitive agent.
Step 2 Rearrange the agents. That is, place the sensitive agent in the first position of the matching lineup.
Step 3 Execute COPE on the rearranged agents.

To identify the sensitive agent, FASTCOPE samples the first Step of each agent for the branch and reads its width. The Step from the sensitive agent will have narrower width than the insensitive agents.

4.3 NEWCOPE Algorithm

Section 5.1 reveals that FASTCOPE can reduce the computational time required to locate the optimal contract by 60%. This is a very big improvement. But, the requirements that , in a branch only one agent can be sensitive can be an unrealistic one. There are many situations where there is a clear difference in sensitivity between the agents but applying FASTCOPE could produce in erroneous results.

One such case is shown in Figure 8. In this case, even though, basically Agent k is an insensitive agent, it has one narrow width step. As a result applying FASTCOPE algorithm would not locate the optimal contract correctly. Another case is the one shown in Figure 9 . For the branch shown there are two agents that are sensitive. As a result again applying FASTCOPE would not locate the optimal contract correctly.

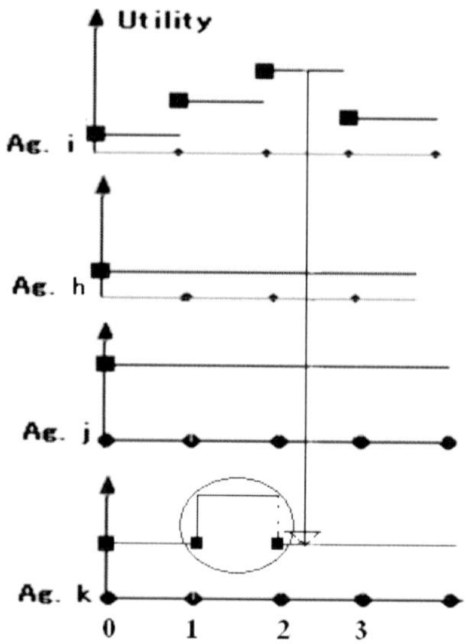

Fig. 8 An insensitive agent has Abnormally narrow width Step

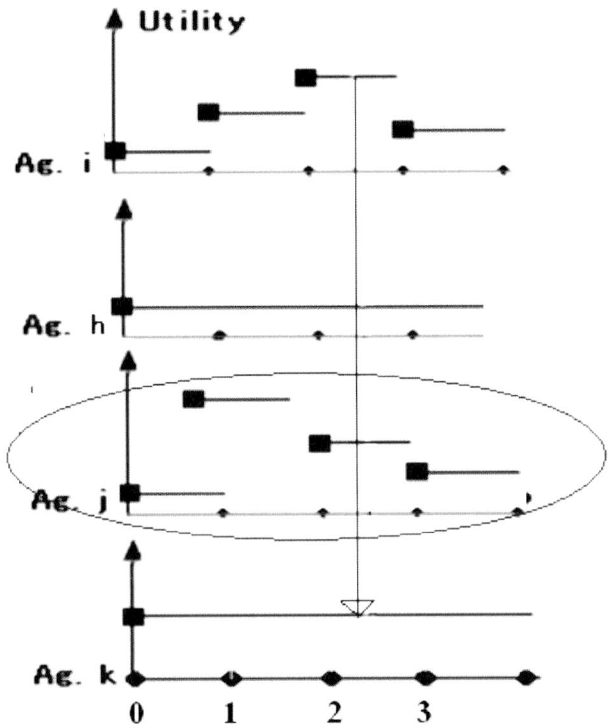

Fig. 9 Two sensitive agents

One way to exploit sensitiveness of agents while ensuring correctness of results is by applying EM on agents sorted in increasing order of their sensitiveness. We call this NEWCOPE algorithm. The reason this algorithm can potentially be efficient can be understood by considering the total number of Step matching required during EM.

When EM starts with the sensitive agent , then for each subsequent agent encountered there will be number of matchings at least equal to the number of Steps the sensitive agent has. And hence the total number of matchings is high (see Figure 10).

On the other hand when EM starts with the insensitive agent and make the sensitive agent the last on the matching line up. The number of matchings with every insensitive agent will be just one(ideal condition). The only time the number of matching is high is when EM reaches the sensitive agent (see Figure 11).

Theoretically the NEWCOPE algorithm certainly improves EM. But in practice its effectiveness can only be verified by experiments. Such experimental result is shown in Sec. 5.2.

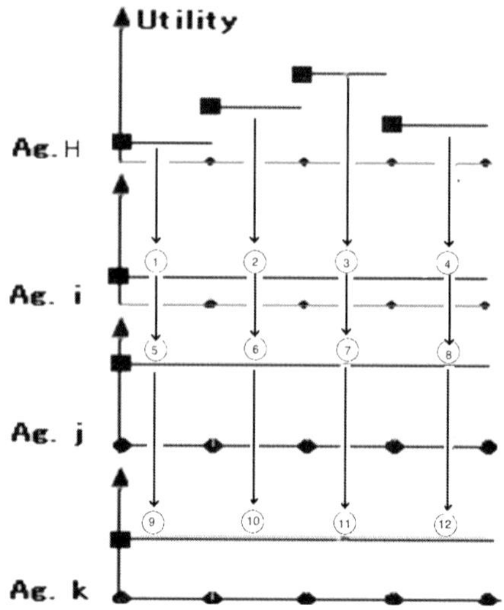

Fig. 10 Number of matching when EM starts with the sensitive agent

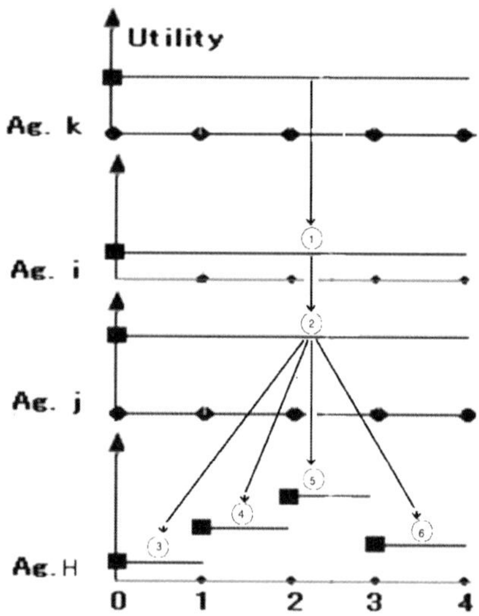

Fig. 11 Number of matching when EM starts with the insensitive agents

5 Experimental Results and Analysis

5.1 *EM vs. Cope vs. Fastcope*

We compared the efficiency of EM, COPE and FASTCOPE experimentally. The result is shown in Figure 12. As expected COPE and FASTCOPE have higher efficiency than EM. COPE (20%) means, 20 percent of the branches satisfy the COPE condition. The rest violate it by not having the first agent as the sensitive one. When COPE is applied on branches that do not satisfy the condition, it makes no efficiency improvement. FASTCOPE rearranges the agents and applies COPE to compute the optimal contract for the branch. The experiments were done at sensitivity ratios of 1:1000, 1:100, 1:10 and 1:5. For example sensitivity ratio of 1:5 means, the entire contract space is divided into branches that contain 5 contracts each. In a branch only one agent is sensitive and it will have 5 narrow width Steps. Each of the remaining agents will have one wide Step. When the total number of the contracts in the negotiation is 10000, there will be 10000/5 , 2000 branches. In figure Figure 8, for each algorithm, the average of the running time costs of the algorithm at the four sensitivity ratios is shown. The number of agents in the negotiation was 4.

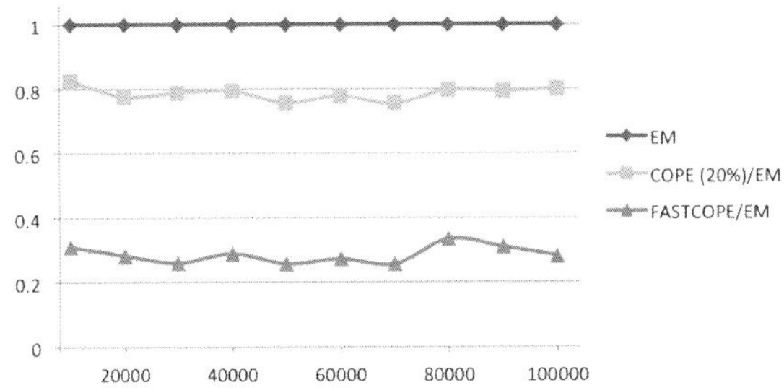

Fig. 12 EM vs COPE vs FASTCOPE

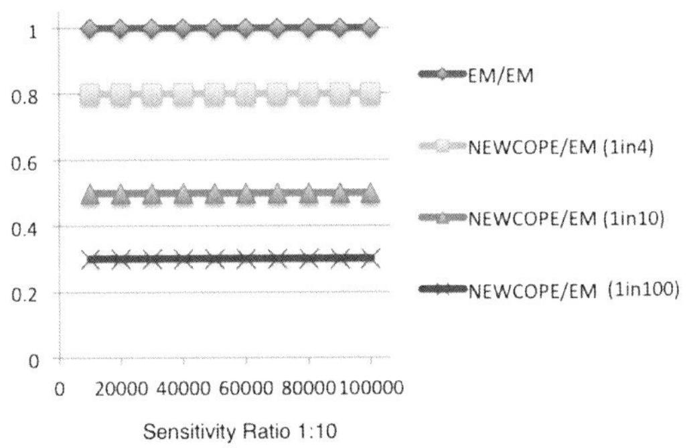

Fig. 13 EM vs NEWCOPE

5.2 *Em vs. Newcope*

Figure 13 shows the experimental result of comparison of execution time of EM and NEWCOPE. In the figue 1inX, means; there are X number of agents and only one of them is sensitive agent. Therefore, for example, 1in10 means;there are ten agents and one of them is sensitive agent.

The sensitivity ratio used in the experiments is 1:10. This means, in a brach there are ten contracts. Moreover, in a branch, while the sensitive agent has ten Steps, insensitive agents have only one Step.

As expected NEWCOPE reduced the computation time required to locate the optimal contract. However, significant improvement (more than 50%) was found only after the number of agents is more than ten. Obviously it is much more preferable if the improvement was significant for even lesser number of agents. But still this result could still be acceptable in domains where the total number of agents is high, and the sensitive agents are only 1-10 percent of the total population.

6 Conclusion and Future Works

A preliminary work for designing efficient algorithm that computes the optimal contract correctly for agents that use the constraints based utility space model was reported. The integer interval was identified to be the basic building unit of the model, and it was used to define the single issue version of it. It was argued that , the agents that use this model can be classified to two extreme kinds:sensitive and insensitive. COPE; an algorithm that computes the optimal contract for a branch correctly and efficiently when the first agent is sensitive and the others are insensitive is proposed. FASTCOPE extends COPE by relaxing the requirement that the sensitive agent has to be the first agent in the matching lineup.

Although FASTCOPE is efficient it imposes stringent requirements on the utility space of agents and hence could fail if any of the requirements is even slightly not satisfied. The NEWCOPE algorithm insures the correctness of results while exploiting the agents difference in their sensitivity. NEWCOPE first sorts the agents in increasing order sensitivity and applies EM.

As a future work, we aim to evaluate NEWCOPE with more than one sensitive agent in the population. For example two or three sensitive agents in negotiation among ten agents. Also we want to investigate on how to allow agents to independently branch their utility space. That is handling the case where the end points of the branches from each agent are not exactly the same (overlap).

Another future work is to extend the algorithm developed for the single issue version of the model to work for multiple issue version of it.

References

1. Fujita, K., Ito, T., Klein, M., Hattori, H.: An approach to implementing a threshold adjusting mechanism in very complex negotiations a preliminary result. In: KICSS, pp. 185–192 (2007)
2. Fujita, K., Ito, T., Klein, M.: Effects of revealed area based selection method for representative based protocol. In: ACAN (2008)

3. Hattori, H., Klein, M., Ito, T.: Using iterative narrowing to enable multi-party negotiations with multiple interdependent issues. In: AAMAS, pp. 1043–1045 (2007)
4. Ito, T., Hattori, H., Klein, M.: Multi-issue negotiation protocol for agents exploring nonlinear utility spaces. In: IJCAI, pp. 1347–1352 (2007)
5. Marsa-Maestre, I., Lopez-Carmona, M.A., de la Hoz, E.: Effective bidding and deal identification for negotiations in highly nonlinear scenarios. In: AAMAS, pp. 1057–1064 (2009)

Agreement among Agents Based on Decisional Structures and Its Application to Group Formation

Rafik Hadfi and Takayuki Ito

Abstract. There has been an increasing interest in automated negotiation systems for their capabilities in reaching an agreement through negotiation among autonomous software agents. In real life problems, a negotiated contract consists of multiple and interdependent issues, and therefore makes the negotiation more complex. In this paper, we propose to define a set of similarity measures used to compare the agents' constraints, utilities as well as the certainties over their possible outcomes. Precisely, we define a decision value-structure which gives a reasonable condition under which agents having similar decision structures can form a group. We think that a collaborative approach is an efficient way to reason about agents having complex decisional settings, but show similarities in their constraints, preferences or beliefs. Agents will tend to collaborate with agents having the same decisional settings instead of acting selfishly in a highly complex and competitive environment. Therefore, formed groups will benefit from the cooperation of its members by satisfying their constraints as well as maximizing their payoffs. Under such criterion, the agents can reach an agreement point more optimally and in a collaborative way. Experiments have been performed to test the existence of the decision value-structure as well as its capability to describe an agent's decision structure. Moreover, the decision value-structure was used for group formation based on measuring the agents similarities.

1 Introduction

Automated negotiation is a process by which a group of autonomous agents interact to achieve their objectives. The agents will attempt to reach an agreement and satisfy their contradictory demands through a bargaining process. In an agent-mediated system, an important aspect of the solution is the way in which the agents negotiate

Rafik Hadfi · Takayuki Ito
School of Techno-Business Administration, Nagoya Institute of Technology
e-mail: rafik@itolab.nitech.ac.jp, ito.takayuki@nitech.ac.jp

T. Ito et al. (Eds.): Complex Automated Negotiations, SCI 435, pp. 55–73.
springerlink.com

to propose contracts to each other, and under specific requirements and constraints. In real life situations, agents have to take into consideration multiple attributes simultaneously during the bargaining process, such as the quality, quantity, delivery time, etc. ([7]). In this paper, we propose to define a new approach to tackle the complexity of utilities with interdependent attributes by providing a new model for multi-attribute utility representation, which takes into consideration the possible interdependencies between attributes. In real world, we believe that people who have similar settings could reach an agreement more smoothly. In this paper, we propose also a new criterion for potential consensus under a number of assumptions, related to the decisional structure of the agent, defined as a Constraint-Utility-Belief space. In fact, adopting a cooperative behavior during the negotiation process may improve the performance of the individual agents, as well as the overall behavior of the system they form, by achieving their own goals as a joint decision [6]. To put this straightforward, we assume that our model is based on the following assumptions. In real life, we believe that people who have similar beliefs (certainties) relative to a specific situation, as well as the same preferences (utilities) over the same common outcomes, could reach a reasonable agreement more optimally and smoothly, than if they had different certainties or preferences over different outcomes. To support this claim, we first describe the different aspects of the decisional structure of an agent as a Constraint-Utility-Belief space. Most importantly, we define a unique decision value-structure for each agent, which gives a reasonable criterion, under which agents' decisional structures can be compared. We point out that in the case of similar decision value-structures, the agents can form groups, as an initial step before making coalitions which satisfy their constraints and maximizes their payoffs. Therefore, the agents can reach an agreement point more efficiently and in a collaborative way. We argue that the advantage of such approach is that the agents having strongly different decisional structures *i.e.* different decisional value-structures, do not need to cooperate. Instead, they can find agents having similar settings, and form groups. To this end, and in the case of multi-attribute negotiation, we must define the main components needed by an agent to make decisions.

There have been several works in the context of multi-attribute negotiation for its importance in commerce as well as in social interactions. Different approaches and methods were proposed to analyze multi-attribute utilities for contracts construction. [12] presented the notion of convex dependence between the attributes as a way to decompose utility functions. [9] proposed an approach based on utility graphs for negotiation with multiple binary issues. [2] proposed also a model inspired from Bayesian and Markov models, through a probabilistic analogy while representing multi-attribute utilities. The same idea was firstly introduced by [11] through the notion of utility distribution, in which utilities have the structure of probabilities. Most importantly, a symmetric structure that includes both probability distributions and utility distributions was developed. In another work by [8], a similar concept was introduced by the notion of Expected Utility Networks which includes both utilities and probabilities. [3] proposed a model which takes into consideration the uncertainties over the utility functions by considering a person's utility function as a random variable, with a density function over the possible outcomes.

The remainder of the paper is structured as follows. Section 2 provides a formal definition of our model based on the notion of Decisional Structure of an agent with all its components. Section 3 describes a method used by the agent to construct his proposals or contracts, based on his decisional structure. In section 4, we elaborate a possible usage of the decisional structure as a group formation criterion through a set of similarity metrics. In section 5, we generalize the use of those metrics by the Decisional Value-Structure function as a method to compare agents' decisional structures. The experiments and the analysis of the model are described in section 6. In section 7 we present the conclusions and outline the future work.

2 Decisional Structure

In the following section we will provide an overview of our theoretical model used for the representation of an agent's decisional structure. In fact, by decisional structure, we refer to the overall settings or information used by the decision maker *i.e.* the agent, to elaborate his strategies and make his decisions. In other words, the decisional structure of an agent can be considered as the decision space of the agent representing all his possibilities. Therefore, we will initially focus on a microscopic representation of an agent i regardless from his environment or the other agents. The macroscopic view will be developed in the next sections in the case of group formation. An agent i will define a unique tuple (1) representing his decisional structure.

$$Agent\ i \longmapsto (G_i, U_i, B_i) \tag{1}$$

This tuple will be characterized by the attributes and constraints of the agent i, represented by a Directed Acyclic Graph G_i [2] . The preferences of the agents will be represented by the utilities U_i of the agent. The agent's beliefs or certainties will be represented by the probability distributions B_i. The tuple can be described in the equations (2).

$$G_i = (V_i, E_i) \tag{2a}$$

$$V_i = \{v^i_j \sim a^i_j\}^n_{j=1},\ a^i_j = (x_1, \ldots x_{m_j}) \tag{2b}$$

$$E_i = V_i \times V_i = \ \{d_j\}^{m_d}_{j=1} \tag{2c}$$

$$U_i = \ \{u^i_j\}^n_{j=1} \tag{2d}$$

$$B_i = \ell_i = \ \{\ell^i_j\}^n_{j=1} \tag{2e}$$

$$= \{\ \ell^i_j \left[p_{i,j,1} : x_{i,j,1},\ \ldots p_{i,j,m} : x_{i,j,m_j}\right]\ \}^n_{j=1} \tag{2f}$$

The static structure of the agent in (2a), defines the attributes (2b) and the dependencies (2c) between them, represented as a Directed Acyclic Graph G_i. In (2b), each vertexv^i_j of the graph corresponds to an attribute a^i_j *i.e.*. An attribute a^i_j is defined as a vector of the possible values that can be taken. In the discrete case a^i_j (2b) and in the continuous case $a^i_j \in [x_1, x_{m_j}]$. In (2c), constraints are represented by the edges $\{d_j\}_{j=1}^{m_d} \subset G_i$, and connect the vertices representing dependent attributes. But, it can be used to compute the utilities by mirroring the same dependence structure as a conditional dependence between the utilities [11]. This dependence structure could be updated dynamically during a negotiation process when the agents are collaborative. In (2d), utility functions U_i of the agent i represented as a function-vector $\{u^i_j\}_{j=1}^n$. In our model, we assume that the decision maker *i.e.* the agent follows the axioms of normative utility functions ($\sum_j u^i_j = 1$) [13]. Furthermore, we assume that the used utility functions have the properties of non-satiation ($u'^i_j(x) > 0$) and risk aversion ($u''^i_j(x) < 0$) [5]. Each utility function u^i_j is defined over a domain D_j related to the possible values taken by the attribute a_j as in (3).

$$u^i_j : D_j \rightarrow [0, 1] \tag{3}$$

Another important aspect of our utility functions is that they are defined in term of dependencies as conditional utilities, and therefore embody the notions of conditional probabilities and probability independence [11]. In our model, we use this representation for the computation of the utilities in respect to the functional dependencies. We refer the reader to the work proposed in [2] and related to conditional utilities and the conditional independence. In (2e), the belief or the certainty structure B_i of an agent i characterized by the lotteries $\{\ell^i_j\}_{j=1}^n$ (2f) where each lottery ℓ^i_j is associated to the attribute a^i_j, according to the probability distribution $p_{i,j}$ over the outcomes $x_{i,j,k} \in a^i_j$ with $\sum_{k=1}^{n_j} p_{i,j,k} = 1$. The lotteries of an agent i over the set of attributes a^i_j can be represented by the lottery (4).

$$\ell^i_j \, [p_{i,j,1} : x_{i,j,1}, \, \ldots, p_{i,j,n} : x_{i,j,n}] \tag{4}$$

The probabilities $p_{i,j}$ are the subjective probabilities [1] of the agent i and represent his certainties about the possible outcomes. Each probability associated to an attribute, can be seen as a random variable over the possible values of an attribute [3].

3 Utility Maximization

3.1 Contract Representation

An agent i will represent a contract $\mathbf{C^i}$ as a vector of attributes $\mathbf{C^i} = (\, a^i_1, \ldots, a^i_j, \ldots, a^i_n\,)$, where each attribute corresponds to a vertex $v^i_j \in V_i$ as we mentioned in (2b). Therefore, finding the optimal contract $\mathbf{C}^*$ having the highest utility among the contracts $\mathbf{C}^{i \in N}$, corresponds to solving the objective function (5) [4].

$$\mathbf{C}^* = \arg\max_{\mathbf{C}} \sum_{i \in \mathbb{N}} u_i(\mathbf{C_i}) \tag{5}$$

However, we assume the existence of a number of constraints, describing the relations or interdependencies (2c) between the attributes [2] . In other words, to compute the utility of a single attribute, we must take into consideration the other attributes. Meanwhile, we will associate a specific utility function u_i to each attribute a_i, with i as an attribute index. The overall utility of a contract $\mathbf{C}$ can be represented as in (6).

$$u(\mathbf{C}) = \sum_{a_i \in \mathbf{C}} u_i(a_i/\{a_{j \neq i}\}) \tag{6}$$

It is obvious that none of the overall attributes are needed to compute the utility of a single attribute. It means that based on a graphical representation of the interdependencies (2c), we will only use the connected attributes. The edges d_i representing the constraints or dependencies between attributes. Since the dependencies will exist only between the connected vertices, each vertex a_i will depend on its parent vertices giving the equation (7).

$$u(\mathbf{C}) = \sum_{a_i \in \mathbf{C}} u_i(a_i/\pi(a_i)) \tag{7}$$

Where $\pi(a_i)$ is the set of all the parents of the vertex a_i. This representation means that in order to compute the utility of the attribute a_i we need to use the attributes $\pi(a_i)$ and their corresponding utility functions. Therefore, the objective function (5) can be written as $\mathbf{C}^* = \arg\max_{\mathbf{C}} u(\mathbf{C})$. The final equation is described as in (8)

$$\mathbf{C}^* = \arg\max_{\mathbf{C}} \sum_{a_i \in \mathbf{C}} u_i(a_i/\pi(a_i)) \tag{8}$$

3.2 Example of Contract Construction

Suppose we are dealing with contracts with a number of attributes equal to 7. The goal is to find the optimal contract $\mathbf{C}^*$ satisfying the interdependencies between the attributes. Each agent will organize his attributes and constraints in a specific way defined by the Directed Acyclic Graph in Figure 1.

As we can see in Figure 1, the DAG will represent the contract from a static viewpoint *i.e.* the structure and the interdependencies between the attributes. Moreover, a utility function u_i has to be associated with each vertex v_i, in order to compute the utility of the corresponding attribute a_i. Based on the graph in Figure 1, the interdependency relations between attributes will yield the same dependencies among the utility functions as shown in Table 1. In the concrete case, an attribute a_j can have different values and therefore will be represented by a vector $a_i = \{x_j \in D_i\}_{j=1}^{m_j}$ Maximizing an utility function u_i is finding the value $x^* \in D_i$ representing the maximum of u_i as in (9).

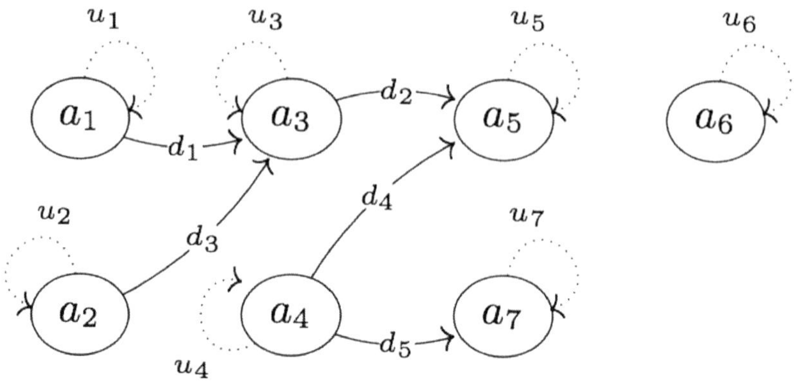

Fig. 1 Proximity of the designed vectors

Table 1 Conditional Utility functions

Utility u_i	Conditional Utility $u_i/\{u_j\}_{j=1}^{7}$
u_1	u_1
u_2	u_2
u_3	$u_3/\{u_1,u_2\}$
u_4	u_4
u_5	$u_5/\{u_3,u_4\}$
u_6	u_6
u_7	$u_7/\{u_4\}$

$$u_i(x^*) \geq u_i(x_k)\ \forall k \in [1, m_j] \tag{9}$$

Thus, we are interested in maximizing the sum of the increasing functions U_i. Therefore, the optimal contract can be written as a vector $\mathbf{C}^* = (\, a_1^*, \ldots, a_i^*, \ldots, a_n^*\,)$, where a_i^* is the maxima of u_i. The optimal contract's utility is computed according to the equation (10).

$$u(\mathbf{C}^*) = \sum_{i \in \mathbb{N}} u_i(a_i^*/\pi(a_i^*)) \tag{10}$$

3.3 *Agent's Optimal Contract*

The algorithm **Optimal_Contract** is used to find the optimal contract based on the attributes (2b), the utilities (2d), and the interdependencies among the attributes (2c).

Based on our example in Figure 1, the vertices a_i will be sorted according to the number of parents *i.e.* the in-degree $deg^-(a_i)$, which will describe the number of constraints of the related attribute.

Algorithm: Optimal_Contract
Input: DAG G_i of the Agent i
Output: Optimal Contract C^*

begin
  Topologic ordering of a_i according to $\pi(a_i)$;
  for $k \leftarrow |\pi(a_i)|_{min}$ **to** $|\pi(a_i)|_{max}$ **do**
    foreach a_i *satisfying* $|\pi(a_i)| = k$ **do**
      Find a_i^* satisfying $u_i(a_i^*) \geq u_i(x_j),\ j \in [1, m_i],\ x_j \in D_i$;
    end
  end
  $C^* \leftarrow (\ a_1^*, a_2^*, a_3^*, \ldots, a_i^*, \ldots, a_n^*\ )$;
  return C^*
end

Fig. 2 Optimal contracts finding Algorithm

An attribute a_i with $deg^-(a_i) = 0$ is called a *free attribute*, as the corresponding utility is computed only by using the attribute a_i's utility function u_i without any reference to other utility functions or other attributes. Similarly, an attribute with $deg^+(a_i) > 0$ is a *non-free attribute* or *dependent* and is subject to $deg^+(a_i)$ constraints. The topological sort of the attributes a_i within G_i is based on the $deg^-(a_i)$.

4 Group Formation

4.1 Group Formation Metrics

The nonlinearity and the complexity of the agents preferences is basically due to the different constraints they are trying to satisfy, as well as their utilities and the way probabilities are assigned. Generally, our approach tends to capture and analyze the similarities between the agents constraints, utilities and beliefs. Being part of the same group means that all its members have close constraints, utilities and certainties. Therefore, it is important to define the similarity functions, to be able to compare between two agents' decisional spaces and decide whether they can be part of the same group or not.

4.2 Metric Related to the Graph

We define the measure *sim* as the degree of similarity between two graphs G_1 and G_2. In other words, how much the agents having graphs G_1 and G_2 share constraints and how close they are in term of vertices and edges. The similarity measure is calculated by multiplying the *Jaccard* indexes relative to the vertices and the edges sets.

This similarity measure can be defined by (11).

$$sim : G \times G \rightarrow [0,1] \tag{11a}$$

$$sim(G_1,G_2) = J_V(V_1,V_2) \times J_E(E_1,E_2) \tag{11b}$$

$$= \frac{|V_1 \cap V_2|}{|V_1 \cup V_2|} \times \frac{|E_1 \cap E_2|}{|E_1 \cup E_2|} \tag{11c}$$

The extreme value $sim(G_1,G_2) = 0$ means that the agent 1 and the agent 2 do not have the same attributes nor share common constraints, whereas $sim(G_1,G_2) = 1$ means that they have exactly the same attributes and the same constraints. Therefore, it might be interesting to consider these similarities' measures between agents' DAGs as a way to form groups and maybe think of potential coalitions. Under these hypothesis, each agent i has a vector $SG_i = \{sim(G_i,G_k)\}_{k \neq i}$ containing all the similarity values between his graph G_i and the other agents' graphs G_k. Using this vector, the agent can selected the set of agents having similar structures (attributes, constraints). This can be a first step for a future collaboration between the agents being part of the same group.

4.3 Metric Related to the Utilities

As mentioned in **2.**, the utility functions have the properties of *non-satiation* and *risk aversion* . Under these hypothesis, we assume that the behavior of these functions can be used to compare the utilities of two agents. Let's consider two utility functions $u_i : D_i \rightarrow [0,1]$, $u_j : D_j \rightarrow [0,1]$ and the domain $D = D_i \cap D_j$. If we suppose that u_i and u_j are similar ($u_i \sim u_j$), then (12) holds.

$$u_i \sim u_j \Longrightarrow \forall x \in D, \exists \varepsilon, |u_i(x) - u_j(x)| \leq \varepsilon \tag{12}$$

The main purpose of comparing utility functions is finding a similarity measure enabling us to say whether two agents have the same preferences over the same attributes or not. We can propose a way to compare two agents' utilities by comparing their accumulated wealth for the same attribute x. In this case, we have to consider the utility value as if it was a cumulative distribution function. Comparing two agents' utilities u_i and u_j is comparing their integrations from the last preferred outcome x_{min} up to the outcome x. Therefore (13) holds.

$$u_i \sim u_j \Longrightarrow \int_{x_{min}}^{x} (u_i(x) - u_j(x))\,\mathrm{d}x \simeq 0 \tag{13}$$

The comparison measure of two utility functions u_i and u_j up to an outcome x will be defined as in (14).

$$sim(u_i, u_j) = \int_{x_{min}}^{x} (u_i(x) - u_j(x))\,\mathrm{d}x \tag{14}$$

We notice that both utilities have the same type *i.e.* correspond to the same attribute (domain). Therefore comparing the overall n utilities U_i and U_j of two agents i and j can be determined as in (15).

$$sim(U_i, U_j) = \prod_{k=1}^{n} sim(u_k^i, u_k^j) \tag{15}$$

4.4 Metric Related to Beliefs

The agents have different certainties when it comes to decide about the outcomes and their related preferences. Therefore, we think about a way to compare these certainties defined as lotteries. Two agent i and j will share the same certainties (beliefs) for an attribute a_k, if their respective probability distributions p_k^i and p_k^j over a_k are close or similar. A possible way to consider this similarity is to use the cross entropy. Assuming that for a certain attribute $a_k = (x_1, \ldots x_{m_k})$ and for two lotteries ℓ_k^i and ℓ_k^j relative to two agents i and j, each lottery will correspond respectively to a probability distributions p_k^i and p_k^j over a_k. Therefore, we can define the cross entropy of p_k^i and p_k^j as in (16).

$$sim(p_k^i, p_k^j) = \sum_{l=1}^{m_k} p_k^i(x_l) \log[p_k^j(x_l)] \tag{16}$$

Generally, each agents i has a vector of lotteries ℓ_i over the n attributes and defined as his certainty structure B_i as in (2e) and (2f). We can define a similarity measure comparing two agent's certainty structures B_i and B_j as in (17).

$$sim(B_i, B_j) = \sum_{k=1}^{n} sim(p_k^i, p_k^j) \tag{17}$$

5 Decisional Structure Value Function

After defining the agent's metrics we will focus on how to exploit them in order to satisfy the common constraints as well as the possible similarities between the agents's belief and utilities. For example, the agents sharing the same constraints (same graphs' structure) and having the same beliefs (same probability distributions over the outcomes) could form groups by opening and sharing their utility functions according to a specific strategy. As in (1), the tuple (G_i, U_i, B_i) of an agent i describes his constraints, preferences and beliefs in a way that identifies the agent from the other agents' configurations. However, if the values G_i, U_i and B_i

represent in a unique way their corresponding agent, it is possible to construct a bijective function f which maps each agents tuple (G_i, U_i, B_i) to a unique real value $dsv_i \in [0,1]$ identifying the agent in a unique way. This function can be assimilated to an *Hilbert Space Filling Curve* [10] or can be constructed by a binary expansion of real numbers. This function can be described by the definition (18).

$$f : D_J \times D_U \times D_P \rightarrow [0,1] \tag{18a}$$

$$f(g_i, u_i, p_i) = dsv_i \tag{18b}$$

The domains D_J, D_U and D_P of f are equal to [0,1]. We will develop in the next section the proper use of this function f in the context of group formation and agents clustering. The function f must be injective *i.e.* for two agents i and j having different settings (g_i, u_i, b_i) and (g_j, u_j, b_j) we will have (19).

$$(g_i, u_i, b_i) \neq (g_j, u_j, b_j) \Longrightarrow f(g_i, u_i, b_i) \neq f(g_j, u_j, b_j) \tag{19}$$

It is possible to prove not only the existence of an injection from $[0,1]^3$ to $[0,1]$ but also a bijection. In fact, that bijection exists and it can be proven using the *Cantor-Bernstein-Schroeder* theorem as following:

i. There is an injection g satisfying (20).

$$g : [0,1] \rightarrow [0,1]^3 \tag{20a}$$

$$g(x) = (x,0,0) \tag{20b}$$

ii. It is possible to define an injection $h : [0,1]^3 \rightarrow [0,1]$ given by representing the tuple (x,y,z) in binary and then interlacing the digits before interpreting the result in base 10, yielding the image of (x,y,z). Using binary for the representation of the strings is a way to avoid the 9's with the dual representation in base 10 and therefore, preserving the injection.
iii. Based on *i.* and *ii.* , we can apply the *Cantor-Bernstein-Schroeder* theorem, which states that if there are two injections g and h as in (21a) and (21b),

$$g : A \rightarrow B \tag{21a}$$

$$h : B \rightarrow A \tag{21b}$$

Then there is a bijection f between A and B. Hence, it is possible to find f satisfying the condition (19).

An interesting usage of the function f is in a mediated negotiation where a mediator is gathering bids from the agents and trying to find the optimal contract. In fact, f provides to the mediator a way to group the agents based on their similarities without the need for the agents to open their utility spaces or their constraints. In this situation, the mediator can establish a feedback mechanism to update his constraints according to the settings of the agents. The convergence to the optimal solutions, ensuring social welfare, will be based upon the agents' feedback as well as the initially established mediator's constraints. Each agent i has only to provide the decisional structure value (*dsv*) which can be seen as a fuzzy indicator about the agent's Constraint-Utility-Belief Space ($[0,1]^3$). Once these values are collected, the mediator can analyze and predict the possibilities of consensus reaching and the convergence to final contract. This is done before starting any utility space sampling or any computationally consuming task, used for example in [4]. The main advantage of using the *dsv* is to avoid bidding when the bids are likely to yield a complex and nonlinear utility space. Furthermore, having nonlinear space tends to make the consensus finding process complex, especially when there is a mediator. In fact, the mediator has to collect the bids and explore a highly nonlinear utility space in order to find the Pareto optimal contracts [4]. Instead, we can find an appropriate grouping of the bids based on certain criteria (including similarity measures) defined by the decisional structure values of the agents.

As we mentioned above, f is bijective, as the agents do not need to open their utilities nor their belief nor their constrains. Instead, they can know exactly how close and how similar their decision structures are and hence to decide whether to go for a collaborative strategy or act regardless from the others. The closeness degree between two agents stands upon the monotonicity of f when mapping to [0,1]. The function f can capture enough information that allows a meaningful clustering of agents based on their common interests: Constraints, Utilities, Belief, etc.

6 Experimental Analysis

6.1 Belief Comparison

In the following section we provide an example of metrics related to belief comparison. In fact, we think that certainty comparison is the most important aspect of the decisional structure of an agent, and that any other component comparison could be reduced to a difference computation. In the following experiment we consider two agents 1 and 2 having respectively two belief structures B_1 and B_2 as in (22). We propose to compare those beliefs assuming that the agents are modeling 3-ary contracts with independent real attributes.

$$B_1 = (X_1, X_3, X_5) \tag{22a}$$

$$B_2 = (X_2, X_4, X_6) \tag{22b}$$

As in (22), the beliefs B_1 and B_2 are lotteries, *i.e.* random vectors. The agents' first components X_1 and X_2 are random variables over the first attribute defined on a domain D_1 (respectively X_3, X_4 over D_2, and X_5, X_6 over D_3). The random variables $\{X_i\}_{i=1}^{6}$ are generated from a seed random variable $X_0 \sim \mathcal{N}(0,1)$ with a certain *Pearson* correlation coefficient r_i for each X_i. For each random variable, we will consider a number of observations equal to 1000. Using the coefficients r_i will allow us to tune the random variables in order to ensure some closeness hypothesis between the agents beliefs. For instance, if $(r_1, r_2) \in [0.7, 1]^2$, there will be a strong positive association between X_1 and X_0 (and respectively X_2 and X_0). In the same way, if $r_3 \in [-1,-0.7]$ and $r_6 \in [-0.3,0.3]$, we will say that X_3 has a strong negative association with X_0 and that X_6 has no association with X_0. The use of correlation will reflect the situation where the agents make their supposition based on some available common information or observation. Hence the correlation concept, which can be a positive linear correlation, a negative correlation, or no correlation at all (independence). We provide two cases where the agents make their observations according to two sets of coefficients. In the first example, the correlation coefficients (23) are generated in a way to yield strong correlation *i.e.* falling within $[0.7,1]$ or $[-1,-0.7]$. Moreover, the agents' coefficients corresponding to the same attribute are symmetrical, as it is shown in (23).

$$\forall i \in \{1,3,5\},\ r_i + r_{i+1} \simeq 0 \tag{23}$$

The symmetry reflects the fact that the agent 1 and 2 have opposite beliefs. In our case, the correlation vector is shown in (24).

$$r = (-0.99,\ 0.99,\ 0.81,\ -0.78,\ 0.91,\ -0.78) \tag{24}$$

As we can see in Figure 2, X_1 was generated from X_0 according to a negative linear correlation, while X_3 and X_5 follow a positive correlation with regard to X_0.

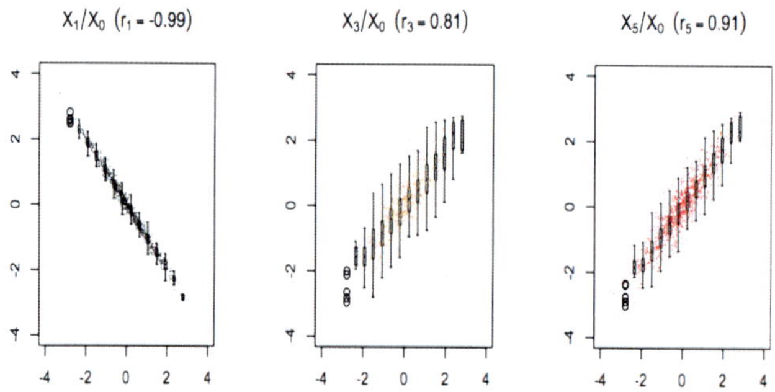

Fig. 3 Belief representation of agent 1 (Case 1)

Furthermore, X_1, X_3 and X_5 are respectively symmetrical to X_2, X_4 and X_6 as shown in Figure 3 and expressed in (23). Once the agent's beliefs B_1 and B_2 are generated, we will be interested in comparing them based on the belief metrics defined in (26). We start by computing the individual divergences between B_1 and B_2 yielding the divergence vectors as in (25).

$$D_{1,2} = (D_{KL}(p_1^1 \| p_2^2),\ D_{KL}(p_3^1 \| p_4^2),\ D_{KL}(p_5^1 \| p_6^2)) \tag{25a}$$
$$= (9.138040,\ 6.494497,\ 7.315457) \tag{25b}$$

Computing the similarity measure between i's and j's beliefs yields (26).

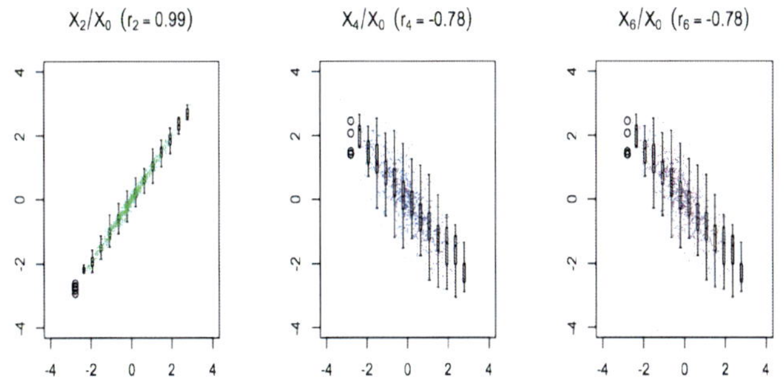

Fig. 4 Belief representation of agent 2 (Case 1)

$$sim(B_1,B_2) = \sqrt{\sum_k [D_{KL}(p_k^1 \| p_k^2)]^2} \tag{26a}$$

$$= \sqrt{D_{KL}^2(p_1^1 \| p_2^2) + D_{KL}^2(p_3^1 \| p_4^2) + D_{KL}^2(p_5^1 \| p_6^2)} \tag{26b}$$

Relatively to the current experiment, the resulting similarity measure is $sim_1(B_1,B_2) = 13.3864$, and represents the distance between the agent 1 and the agent 2 from the belief (uncertainty) perspective. The accuracy and the truthfulness of this result will be analyzed once we get the results of the opposite case *i.e.* the situation where the agents have strongly correlated beliefs. In this case, the coefficients of the same attributes (same component on both B_1 and B_2) will tend to have the same sign and fall within $[0.7,1]$ or $[-1,-0.7]$, as it is expressed in (27).

$$\forall i \in \{1,3,5\},\ r_i - r_{i+1} \simeq 0 \tag{27}$$

Therefore, we will consider another correlation vector described in (28).

$$r = (0.89,\ 0.91,\ -0.82,\ -0.93,\ 0.76,\ 0.89) \tag{28}$$

As we can see in Figure 4 and Figure 5, all the attributes are strongly correlated, whether it is a negative or a positive association.

The resulting difference vector represented in (29) shows the degree of closeness between the attributes X_1 and X_2 (respectively X_3, X_4 and X_5, X_6).

$$D_{1,2} = (0.008321388,\ 0.164517754,\ 0.191820548) \tag{29}$$

Similarly to the previous case, we compute the distance between B_1 and B_2 by computing the norm of $D_{1,2}$, yielding $sim_2(B_1,B_2) = 0.2528$. The correlation hypotheses for both cases could be expressed by comparing the similarity measures found for both situations. In the first case, the agents show high dissimilarity in the way their certainties are defined. But in the second case, they seem to show similarities. This is described by : $sim_2(B_1,B_2) < sim_1(B_1,B_2)$. Therefore, the used similarity

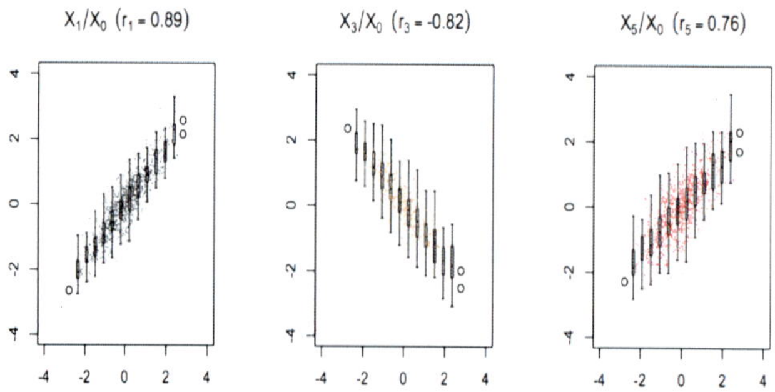

Fig. 5 Belief representation of agent 1 (Case 2)

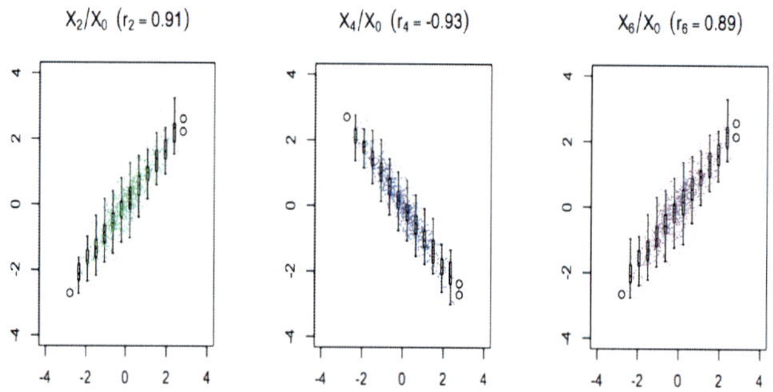

Fig. 6 Belief representation of agent 2 (Case 2)

measure, defined in (26), can be used to compare two agents' beliefs. As a future work, we will develop the case where the agents' beliefs are uncorrelated.

6.2 *Group Formation*

In the following experiments, we provide a method for group formation based on the similarity between the decisional values of the agents. We also provide an application of the decisional structure in the design of vectors called *vectorial design*.

Given the set $C = \{d_i\}_{i=1}^{N}$ of all the decisional structure values (dsv) of the agents, we propose to partition C into k disjoint clusters using the *K-Means* algorithm. Finding the optimal partitioning of C corresponds to finding the k clusters as in (30).

$$C^* = \arg\min_{C} \sum_{i=1}^{k} \sum_{d_j \in C_i} \left\| d_j - \delta_i \right\|^2 \tag{30}$$

Each cluster or group C_i is centered around a specific structure value δ_i which refers to the agent having the decisional structure that is more likely to describe

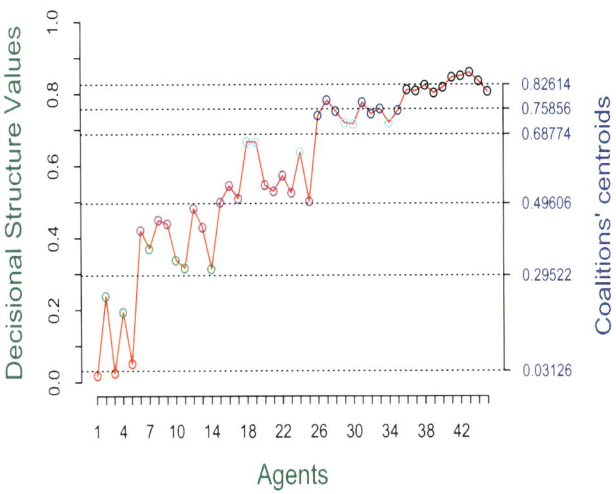

Fig. 7 Agents' *dsv* values

the common features of the group C_i. Figure 6 illustrates a process of grouping 45 agents, based on their decisional structure values (d_j). We propose to partition these agents into 6 groups each of which is characterized by a group centroid δ_i. The resulting groups can be described by their corresponding centroids which are represented in Figure 6 in blue, on the right axis. The decisional structure values δ_i were generated based on the function f defined in **5**., which was applied on the G_i, U_i and B_i variables of the 45 agents. The corresponding DSVs must be unique for each agent. Under such hypothesis, the injectivity of the function f will hold and there will be no risk for collisions *i.e.* two different agents, having different decisional structures but having the same DSV. Based on the original tuples G_i, U_i and B_i, we found that the agents being part of a group (C_j, δ_j) had close constraints, utilities and probabilities. This result was evaluated firstly by comparing the similarities between two agents decisional structure values dsv_i and dsv_j based on the distance $d = |dsv_i - dsv_j|$. Secondly, we measured the distances $d_g = sim(G_i, G_j)$, $d_u = sim(U_i, U_j)$ and $d_b = sim(B_i, B_j)$, defined in **4**. We found that the distance d is related to the distances d_g, d_u and d_b. The result confirms the characteristics of the bijective function f defined in **5**., and its ability to describe uniquely an agent's decisional structure.

In Figure 7, we can see that there is a number of agents grouped around the same dsv value. In this case, the agents 2, 3, 4, 5 and 9 can be grouped into a cluster G based on the assumption that they have common decisional structures. According to this information, and whenever its shared to the overall agents (1 to 10), the agents not being part of G can choose to join this group or not. In case they accept to join, it is probable that they should start adapting and updating their constraints, preferences and beliefs similarity to the initial agents of G.

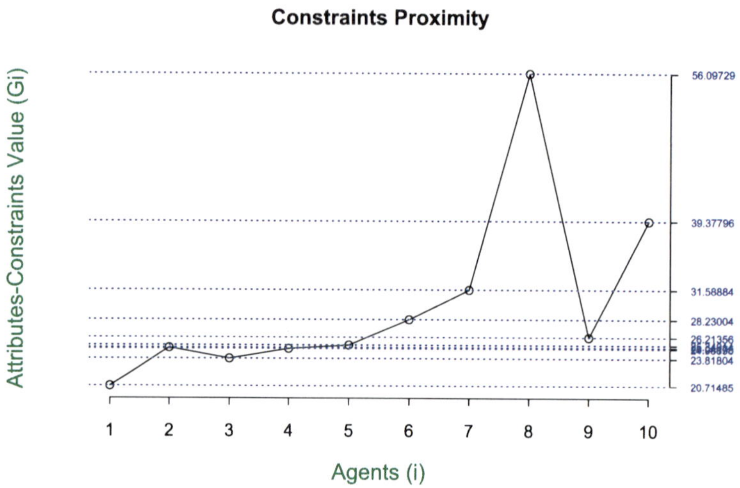

Fig. 8 Dominant Group

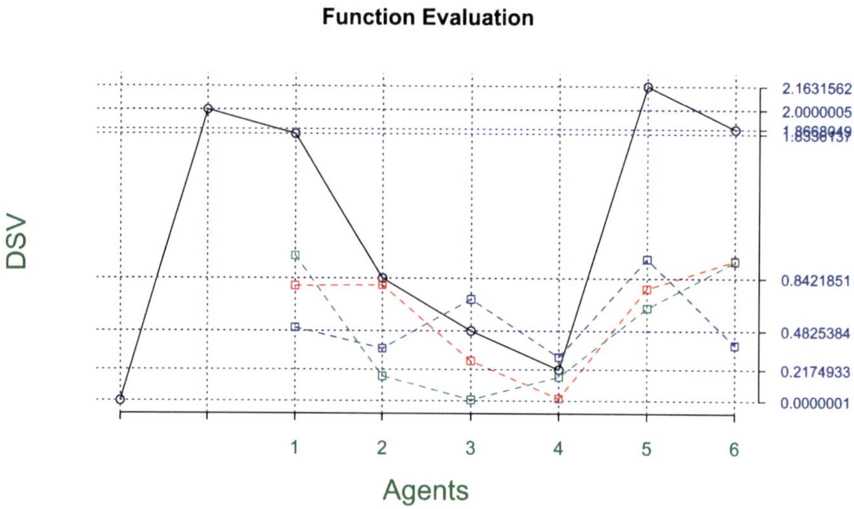

Fig. 9 DSVs comparison

Generally, The decisional value structures are constructed based on the graphical constraints, utilities and beliefs. As we can see in Figure 8, the red curve represents the graphical constraints values, the utilities are represented by the green curve, and the blue values represent the beliefs. The overall similarity is represented by the black curve. For example, we can see that the agent 1 and the agent 6 have close DSVs, and this can be seen based on the closeness in the red, green and blue curves *i.e.* the graphical constraints points, utilities and belief points.

Table 2 Agents' vectors values

Agents	x_1	x_2	*Values*
A_1	0.12	0.96	0.05683
A_2	1.87	1.83	0.68083
A_3	1.34	1.45	0.38637
A_4	1.41	1.57	0.44097
A_5	2.32	2.92	1.36735
A_6	2.39	3.01	1.4523

A concrete application of such method of comparison is the case of *vectorial design*, where a user designs graphically a vector. A vector can represent an object, a product, or more generally a multi-attribute contract. As an example, 6 agents are designing 6 different vectors. For the sake of simplicity, we can think about the vector as a 2-points vector with components x_1 and x_2. In Table 2 we can see that for each two values x_1 and x_2 we can represent the design vector by a unique value, locating the agents design in the overall designed vectors. This will give an idea about the degree of closeness between the designed vectors. The degree of closeness of the agents's vectors can be provided as a shared information to the overall agents while they are designing their vector. In fact, sharing such information dynamically and in real time can give the agents an idea on how their vectors are located in the group, and how to change their vector accordingly. This information can be represented as in Figure 9, and is available to each agent. On the x axis, we have the agents's indexes from 1 to 6 represented by 6 bars, and on the y axis we represent their corresponding values. Whenever an agent changes his vector, the representation in Figure 9 will change accordingly. Such method of collaborative design will give the agents the possibility to orient their design based on the overall group's preferences, ensuring social welfare. It is possible to extend the simple vector

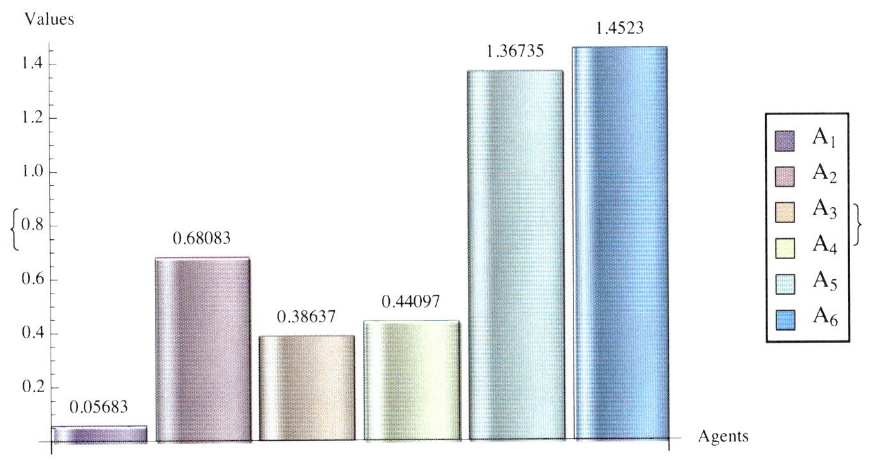

Fig. 10 Proximity of the designed vectors

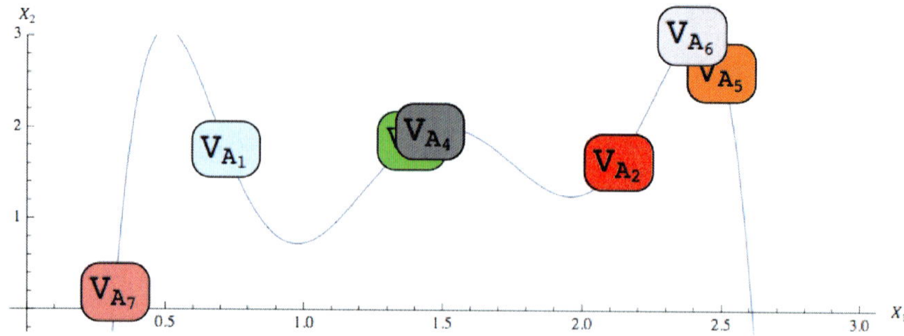

Fig. 11 Vectors representation

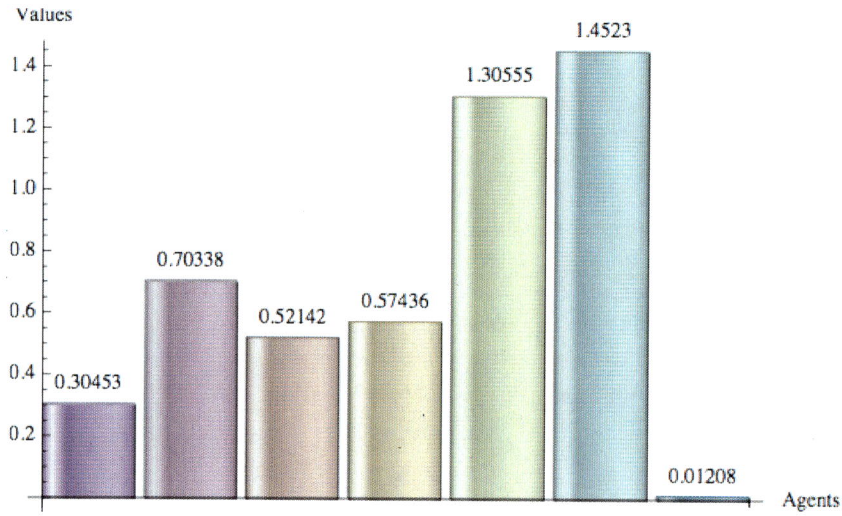

Fig. 12 Decisional Values representation

represented by x_1 and x_2 to a more complex vector. Another example of vectorial design is represented in Figure 10 where 7 agents are designing 7 vectors. At different times, each agent A_i will provide a vector $V_{A_i} = (X_{i1}, X_{i2})$, where X_{ij} are real values. During the design process, each agent A_j can visualize the similarities between his design and the other agents $A_{k \neq j}$ as in Figure 11. Therefore A_j can update his vector according to the evolution of the other agents' designs.

The represented values in Figure 10 correspond to the designed vectors represented in Figure 11. We can can see that the vectors V_{A_6} and V_{A_5} are graphically close in Figure 10, therefore their corresponding values in Figure 11 will be also close (1.30555 and 1.4523). The same comparisons can be done to the other vectors, allowing the agents to see the likelihood and the convergences of the global design.

7 Conclusion

The contributions of this paper are two-fold. On the one hand, we proposed a theoretical model to reason about multi-attribute contracts representation taking into consideration the attributes' interdependencies. On the other hand, we provided the notion of decisional structure value as a main criterion for agents' decisional settings comparison. The defined structure-value captures the main similarities between the agents' decisional settings. We have shown that it is possible to represent such decisional setting as a Constraints-Utilities-Belief space. Furthermore, we provided an example of usage of such value in the case of group formation based on the degree of similarity between the agent's decisional spaces.
As a future work, we would like to consider the performances of the method used to generate the decisional structure value. Moreover, we would like to elaborate a complete negotiation process, by defining a concrete protocol based on the formed groups. For example, we can develop the case where the agents being part of the same group can open and share their utility functions.

References

1. Anscombe, F.J., Aumann, R.J.: A definition of subjective probability. The Annals of Mathematical Statistics 34(1), 199–205 (1963)
2. Brafman, R.I., Engel, Y.: Directional Decomposition of Multiattribute Utility Functions. In: Rossi, F., Tsoukias, A. (eds.) ADT 2009. LNCS, vol. 5783, pp. 192–202. Springer, Heidelberg (2009)
3. Chajewska, U., Koller, D.: Utilities as random variables: Density estimation and structure discovery. In: Proceedings of the Sixteenth Annual Conference on Uncertainty in Artificial Intelligence, UAI 2000, pp. 63–71 (2000)
4. Ito, T., Hattori, H., Klein, M.: Multi-issue negotiation protocol for agents: Exploring nonlinear utility spaces. In: Proceedings of the 20th International Joint Conference on Artificial Intelligence, IJCAI 2007, pp. 1347–1352 (2007)
5. Keeney, R.L., Raiffa, H.: Decisions with multiple objectives. Cambridge University Press (1993)
6. Kraus, S., Schechter, O.: Strategic-negotiation for sharing a resource between two agents. Computational Intelligence 19, 9–41 (2003)
7. Lai, G., Sycara, K.: A generic framework for automated multi-attribute negotiation. Group Decision and Negotiation 18, 169–187 (2009)
8. Mura, P.L., Shoham, Y.: Expected utility networks. In: Proceedings of the Conference on Uncertainty in Artificial Intelligence (1999)
9. Robu, V., Somefun, D.J.A., Poutre, J.L.: Modeling complex multi-issue negotiations using utility graphs. In: Proceedings of the 4th International Joint Conference on Autonomous Agents and Multi-Agent Systems, AAMAS 2005, pp. 280–287 (2005)
10. Sagan, H.: Space-filling curves. Springer, New York (1994)
11. Shoham, Y.: A symmetric view of utilities and probabilities. In: Proceedings of the Fifteenth International Joint Conference on Artificial Intelligence, IJCAI 1997, pp. 1324–1329. Morgan Kaufmann Publishers Inc., San Francisco (1997)
12. Tamura, H., Nakamura, Y.: Decompositions of multiattribute utility functions based on convex dependence. Operations Research 31(3), 488–506 (1983)
13. von Neumann, J., Morgenstern, O.: Theory of Games and Economic Behavior, 2nd edn. Princeton University Press, Princeton (1947)

An Adaptive Bilateral Negotiation Model Based on Bayesian Learning

Chao Yu, Fenghui Ren, and Minjie Zhang

Abstract. Endowing the negotiation agent with a learning ability such that a more beneficial agreement might be obtained is increasingly gaining attention in agent negotiation research community. In this paper, we propose a novel bilateral negotiation model based on Bayesian learning to enable self-interested agents to adapt negotiation strategies dynamically during the negotiation process. Specifically, we assume that two agents negotiate over a single issue based on time-dependent tactic. The learning agent has a belief about the probability distribution of its opponent's negotiation parameters (i.e., the deadline and reservation offer). By observing opponent's historical offers and comparing them with the fitted offers derived from a regression analysis, the agent can revise its belief using the Bayesian updating rule and can correspondingly adapt its concession strategy to benefit itself. By being evaluated empirically, this model shows its effectiveness for the agent to learn the possible range of its opponent's private information and alter its concession strategy adaptively, as a result a better negotiation outcome can be achieved.

1 Introduction

Negotiation is a fundamental topic in multi-agent systems because it allows self-interested agents to achieve mutually beneficial agreements and partition resources efficiently and effectively [9]. In recent years, researchers have paid their increasing attention to the integration of learning techniques into agent negotiation [1] [2] [3] [4] [5] [8] [11] [13]. In this type of learning circumstances, agents need adapt themselves to the changes of opponents and/or the environment through learning in order to achieve a satisfactory result. One promising paradigm of involving learning

Chao Yu · Fenghui Ren · Minjie Zhang
School of Computer Science and Software Engineering,
University of Wollongong, Wollongong, NSW, 2522, Australia
e-mail: cy496@uowmail.edu.au,
fren,minjie@uow.edu.au

T. Ito et al. (Eds.): Complex Automated Negotiations, SCI 435, pp. 75–93.
springerlink.com

in negotiation is through opponent modeling to let agents learn the model of their opponent/teammates (such as preferences, private information and capabilities etc.) in the environment so as to make good predications for expected behaviors and to act accordingly to cooperate with the teammates more effectively or take the best advantage of the opponents. However, due to the essence of competition, privacy and uncertainty in real-life negotiation [10], negotiators are always unwilling to reveal their private information (e.g., parameters such as the deadline, reserve price, or strategy profiles) to their opponents in case of being forced to a worse outcome, thus making learning in negotiation a challenging problem.

In current literature, a number of approaches have been developed by employing agents learning methods into negotiation process. Zeng and Sycara proposed an approach based on Bayesian learning to learn the opponent's reserve price [16]. Their approach assumes that agents have priori knowledge about the opponent's bidding strategy. This assumption may not always be true in real-world negotiations. Hindriks and Tykhonov also proposed an approach to discover opponent's information [5] by using Bayesian learning based on the assumptions that 1) agents know the opponent's weights ranking on negotiation issues and 2) all agents' preferences can be modeled by three proposed functions, which may impact the use of this approach in a wide range when these assumptions conflict with the real world situations. Ren and Zhang introduced approaches based on regression analysis to predict the opponent's concession strategy by using the historical offers only [6] [7]. However, their approaches did not give further advice on how to adapt agent self's concession strategy based on the learning results. Brzostowski and Kowalczyk also presented a way to estimate partners' behaviors in different types of agents, based only on the historical offers in the current negotiation [10]. However, the accuracy of classification on partners' types may impact the accuracy of prediction results. The current challenging issues in agent learning during negotiation include (1) how to design a learning method without priori knowledge of the opponent's private information, (2) how to develop an effective learning strategy only based on the historical offers of current negotiation, and (3) how to produce a constructive guidance from learning to adapt agent's negotiation behaviors so as to achieve a better negotiation outcome.

This research attempts to address the above three challenging issues. In this paper, we propose a novel model by combining Bayesian learning and a regression analysis approach to dynamically learn the opponent's negotiation deadline and reservation offer for an adaptive negotiation process. To be more specific, firstly, a negotiation agent defines some regions and evenly initialize the probability of each region. The probability here indicates how likely that the opponent's deadline and reservation offer are located in the corresponding region. By using the predefined regions, the agent can have some estimations on the opponent's negotiation behaviors. Secondly, by using the regression analysis, the differences between the opponent's real negotiation behavior and the agent's estimated results are calculated. The more similar between the opponent's real behavior and an estimated behavior, the more likely that the opponent's real deadline and reservation offer are located in the corresponding region. Thirdly, based on the similarities between the opponent's real behavior and the estimated behaviors, the probabilities assigned to each region will be updated dynamically through Bayesian learning updating rule. Lastly, the agent will propose a countermeasure for each estimated behavior of the opponent, and all countermeasures will be combined based on the likelihood of each estimated behavior. The combined result will be employed by the agent to perform a reasonable reaction. During the negotiation, each region's probability will be dynamically updated and gradually close to the real situation. Thus, the agent will also gradually adapt its negotiation strategy to reach a better negotiation outcome. Our model only uses historical offers in the current negotiation, without requesting prior knowledge about the environment and the opponent.

The remainder of this paper is structured as follows. In Section 2, we recap the general negotiation model, especially the basic principles of the time dependent tactic. The proposed learning model is introduced in detail in Section 3, and in Section 4 empirical evaluation and analysis are presented. The discussion and related work are given in Section 5. Finally the paper is concluded and some directions for future work are outlined in Section 6.

2 A General Negotiation Model

Before laying out our learning model, we give a brief description of a time dependent, bilateral single-issue negotiation model, which is widely used in many applications. Let i ($i \in \{b,s\}$) represent a negotiator, i.e., b for a buyer agent and s for a seller agent. Both agents have an initial price IP_i and reserve price RP_i for the negotiating issue. The interval $[IP_i, RP_i]$ indicates the range of all the possible agreements, and can be normalized in-between $[0,1]$ using a utility function. In this paper, we choose the widely accepted linear utility function [15] shown in Equation 1:

$$u_i(p_i) = \frac{p_i - RP_i}{IP_i - RP_i} \qquad i \in \{b,s\} \tag{1}$$

where p_i is the value of an offer in the range of $[IP_i, RP_i]$.

In time dependent tactic, agent i concedes its utility $u_i(t)$ under the time constraint. At the beginning of negotiation, agent i has its highest utility of 1 for the *initial price*. As the negotiation proceeds on, the utility $u_i(t)$ decreases according to the decision functions [15], which are a family of polynomial functions given by Equation 2.

$$u_i(t) = 1 - (\frac{t}{T_i})^{\beta} \qquad i \in \{b,s\} \tag{2}$$

where T_i is the deadline of agent i and β is the concession parameter. Figure 1 shows three different concession strategies called *Conceder, Boulware, Linear*, respectively, signifying different concession rates in the negotiation process.

- *Conceder*: When $0 < \beta < 1$, the agent decreases its utility quickly at the early stage of negotiation and slowly when the deadline is approaching.
- *Linear*: When $\beta = 1$, the agent's utility decreases at a constant rate throughout the negotiation process.

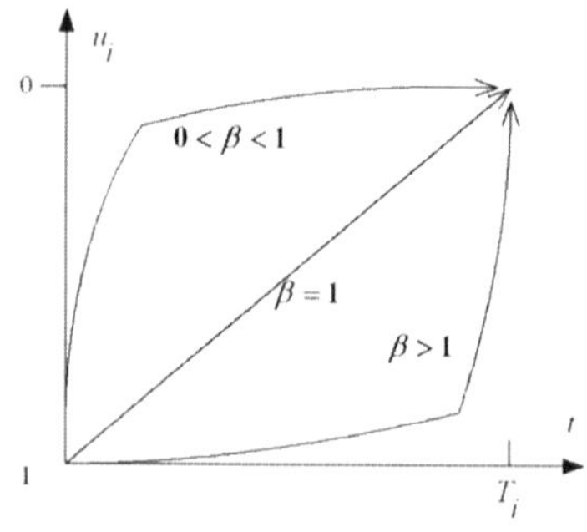

Fig. 1 Three different kinds of concession strategies [15]

- *Boulware*: When $\beta > 1$, it is the contrary of Conceder, which means slow concession at the beginning but quick concession at the late stage of negotiation.

When β is settled, the utility $u_i(t)$ can be computed during the negotiation. As a result, the agent can give a counter offer at time t according to the following offer generating equation.

$$Offer_i(t) = RP_i + u_i(t)(IP_i - RP_i) \qquad i \in \{b, s\} \tag{3}$$

Combining Equation 2 and Equation 3, the offer generating Equation 3 is rewritten as Equation 4.

$$Offer_i(t) = IP_i + (RP_i - IP_i)(\frac{t}{T_i})^{\beta} \qquad i \in \{b, s\} \tag{4}$$

In a non-learning negotiation setting, once an agent sets the value of concession parameter β, the agent will keep this value unchanged through the negotiation process, without any adaptation to the dynamic environment or the revelation of opponent's private information. However, if the agent can learn some useful information from the opponent during the negotiation, it will be able to adapt its original concession strategies and gain more benefits to produce good outcomes for negotiation. In the following section, we will present an adaptive negotiation model using regression analysis and Bayesian learning to enable an agent to alter its concession strategy dynamically, thereby a better outcome will be obtained.

3 An Adaptive Negotiation Model

In this section, an adaptive negotiation model is proposed. This model includes two parts: a learning mechanism and an adaptive concession strategy. Each part will be introduced in detail by Subsections 3.2 and 3.3, respectively. In this paper hereinafter, the discussion is taken from the perspective of the buyer agent unless otherwise specified. However, such a discussion will not lose the generality of our model, i.e. a seller agent can also use our model to learn its opponent's behaviors.

3.1 *Model Description*

As we can see from Equation 4, the parameters of deadline and reserve price are two main factors dominating the negotiation process and outcomes. If agents can obtain the information about these two parameters from the opponent, a better strategy can be employed to increase agents' benefits and/or the negotiation efficiency. Our ami is to model the opponent in terms of these two private information. Before going deep into our model, we firstly give the definitions needed for further illustration.

Definition 0.1. Let x-axis represent negotiation time and y-axis represent the negotiation price. A *detecting region DetReg* is a rectangle in this two-dimensional area to present an estimation of the opponent's deadline and reserve price. This area is defined by a 4-tuple $DetReg = (T^l, T^h, P^l, P^h)$, where T^l, T^h are the estimated lower and upper boundary of the opponent's deadline, and P^l, P^h are the estimated lower and upper boundary of the opponent's reserve price.

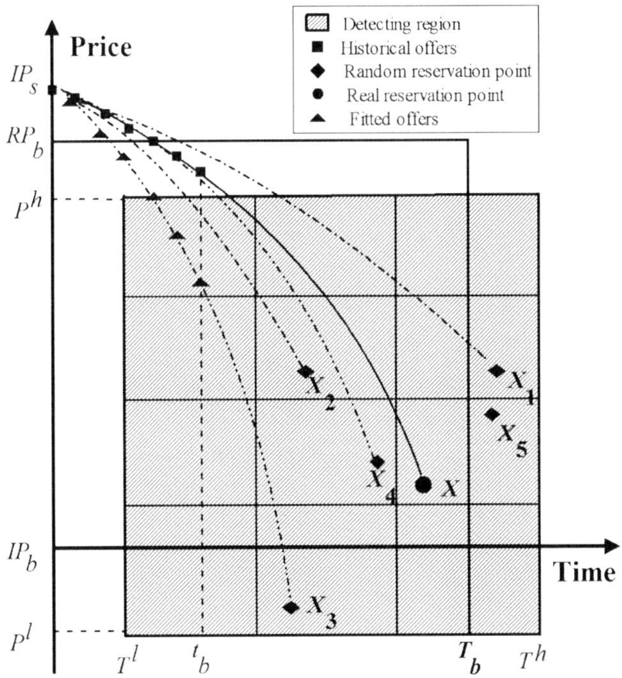

Fig. 2 An example of demonstrating our learning process

As shown in Figure 2, the shadowy area indicates the detecting region for a buyer agent during a learning process based on Definition 1. T_b is buyer's deadline and t_b is the current time in negotiation. IP_b and RP_b represent the buyer's initial price and reserve price, respectively, and IP_s is the seller's initial price. Points appeared in the detection region of the figure will be explained in Definition 3. The lines shown in the figure will be explained in Subsection 3.2 during introducing the learning mechanism.

A buyer agent can initialize the value of each component of *DetReg* according to its estimation about seller's private information. The more precise the estimation is, the smaller the detecting region will be and the buyer can strive for a better result because more errors can be avoided when the buyer agent adapts its concession strategy based on this estimation.

After confirming the detecting region by the buyer agent, this region will be further divided into smaller areas according to $N = (N^t, N^p)$ in which N^t denotes that the detecting region is evenly divided into N^t columns on the x-axis (i.e. time values), and N^p stands for the row number on the y-axis (i.e. price values) in the detecting region. In this way, the detecting region can be divided into a number of smaller blocks, called *detecting cells*. The total number of detecting cells in a detecting region is represented by N_{all} and can be calculated by the formula N_{all}=$N^t \times N^p$. Fig. 2 exemplifies a scenario with $N = (3,4)$ and there are totally 12 detecting cells in the whole detecting region.

Definition 0.2. A *detecting cell* C_i $(i \in 1,2,...,N_{all})$ is a divided block in the detecting region, which can be denoted by a 4-tuple $C_i = (t_i^l, t_i^h, p_i^l, p_i^h)$ where t_i^l, t_i^h are the lower and higher boundaries of time in the cell and p_i^l, p_i^h are the lower and higher boundaries of price in the cell, respectively.

Definition 0.3. A *random reservation point* $X_i(t_i^x, p_i^x)$ is a randomly selected point in each cell C_i, where $t_i^l < t_i^x < t_i^h$ and $p_i^l < p_i^x < p_i^h$.

In Figure 2, points X_1, X_2, X_3, X_4 are several random reservation points in the detecting region and point X is the real reservation point of the opponent. The detecting cell is a region where seller's real reservation point X might be located. That means the real reservation point X might be out of the detecting region in real case. The buyer agent has some belief about the probability distribution of all the detecting cells. The probability of each cell signifies the likelihood that the opponent's real reservation point X might be located in this cell. This belief can be revised more precisely through learning from opponent's historical offers (see Subsection 3.2). Based on this learning result, the agent can adjust its concession strategy adaptively (see Subsection 3.3) to gain more profit over its opponent.

3.2 *The Learning Mechanism*

The purpose of this leaning mechanism is to let the agent revise its belief about the probability distribution of the cells in the detecting region. Because the agent has no knowledge about the opponent, it is hard to determine the precise location of the real reservation point. However the agent can observe its opponent's historical offers to renew the belief about the approximate range of the reservation point. This mechanism consists of two parts, a regress analysis and a Bayesian learning. In regression analysis, (1) an agent chooses a random reservation point in every detecting cell first, based on the belief that this point is the reservation point of the opponent; (2) the agent conducts the regression analysis for all random reservation points corresponding to all detecting cells, respectively; (3) the agent compares the fitted offers on each regression line with opponent's historical offers by the non-linear correlation. By this way, resemblance between the selected random reservation point and the opponent's real reservation point can be calculated. The bigger the non-linear correlation between two lines is, the more alike they will be. This also means that the randomly chosen reservation point has a bigger possibility to be the real reservation point. Then by using Bayesian learning, the agent's belief on the probability distribution will be dynamically updated at every step of the negotiation. The regression analysis and our Bayesian learning method are introduced in the following two subsections, respectively.

3.2.1 Regression Analysis

Before the leaning process, the buyer should initialize *DetReg*, N as well as the probability distribution in each detecting cell, which presents the likelihood that the seller's reservation point is in this cell. When the learning begins, the buyer can do the following steps sequentially.

Step 1: At round t_b, the buyer selects a random reservation point $X_i(t_i^x, p_i^x)$ in each cell C_i of the detecting region;

Step 2: Using each point $X_i(t_i^x, p_i^x)$ chosen in Step 1, the buyer calculates the regression line l_i based on the seller's historical offers $O_{t_b} = \{p_0, p_1, ..., p_{t_b}\}$ until round t_b. Based on Equation 4, the following power regression function is generated to calculate the regression curve.

$$Offer_i(t) = p_0 + (p_i^x - p_0)(\frac{t}{t_i^x})^b \tag{5}$$

where p_0 is the *initial price* of seller. The regression coefficient b is the concession parameter β in the utility function in Equation 4. Then we can calculate coefficient b based on seller's historical offers O_{t_b} by Equation 6 as proposed in [6].

$$b = \frac{\sum_{i=1}^{t_b} t_i^* p_i^*}{\sum_{i=1}^{t_b} t_i^{*2}} \tag{6}$$

where $p_i^* = \ln \frac{p_0 - p_i}{p_0 - p_i^x}$, $t^* = \ln \frac{t}{t_i^x}$. In Figure 2, the solid line is the curve of the seller's historical offers while the dashed line is the regression curve based on each random reservation point.

Step 3: Based on the calculated regression line l_i given by Equation 5 and 6, the buyer can calculate the fitted offers $\hat{O}_{t_b} = \{\hat{p}_0, \hat{p}_1, ..., \hat{p}_{t_b}\}$ at each round.

Step 4: The buyer calculates the non-linear correlation between seller's historical offers O_{t_b} and the fitted offers $\hat{O}_{t_b}$. The coefficient of nonlinear correlation γ can be calculated by Equation 7.

$$\gamma = \frac{\sum_{i=1}^{t_b}(p_i - \bar{p})(\hat{p}_i - \bar{\hat{p}})}{\sqrt{\sum_{i=1}^{t_b}(p_i - \bar{p})^2 \sum_{i=1}^{n}(\hat{p}_i - \bar{\hat{p}})^2}} \tag{7}$$

where $\bar{\hat{p}}$ is the average value of all the fitted offers till time t_b and $\bar{p}$ represents the average value of all the historical offers of the seller. The non-linear correlation γ, where $(0 \leq \gamma \leq 1)$, is a parameter reflecting the non-linear similarity between the fitted offers and the historical offers, which can be used as a criterion to evaluate the resemblance between the random reservation point X_i and seller's real reservation point X. This is an important parameter to be used in Bayesian learning for the belief updating as described in the following section.

3.2.2 Bayesian Learning

In general, Bayesian learning can be used when an agent has a set of hypotheses about its opponent's information. The belief about the probability distribution of these hypotheses can be revised through a posterior probability by observing the outcome of its opponent. In our model, we define the hypothesis space as $H_i, (i \in 1,2,3,...,N_{all})$, where N_{all} is the total cell number in the detecting region. Each hypothesis H_i stands for the assumption that seller's reservation point X is in cell C_i. The prior probability distribution, denoted by $P(H_i), (i \in 1,2,3...N_{all})$, signifies the agent's belief about the hypothesis, that is, how likely the hypothesis fits the real situation. At first, the agent can initialize the probability distribution of the hypotheses based on some public information if available, otherwise a uniform distribution $P(H_i) = 1/N_{all}$ is assigned.

During each round of negotiation t_b, the probability of each hypothesis can be altered by the Bayesian updating rule given in Equation 8.

$$P(H_i|O) = \frac{P(H_i)P(O|H_i)}{\sum_{k=1}^{N_{all}} P(O|H_k)P(H_k)} \tag{8}$$

where the conditional probability $P(O|H_i)$ represents the likelihood that outcome O might happen based on hypothesis H_i. In our learning model, the agent has no information about its opponent, thus the observed outcome O is opponent's historical offers $O_{t_b} = \{p_0, p_1, ..., p_{t_b}\}$. The conditional probability $P(O|H_i)$ thereby means how likely seller's historical offer O_{t_b} can happen based on the hypothesis H_i that seller's real reservation point X is in cell C_i. The posterior probability $P(H_i|O)$ is a renewed belief based on the observed outcome O and at next round, the agent will update the prior probability $P(H_i)$ using the posterior probability $P(H_i|O)$, thus a more precise estimation is achieved by using Equation 8.

To let the Bayesian learning rule work, the most critical problem is how to obtain the conditional probability $P(O|H_i)$. Most approaches using Bayesian learning method usually require a priori knowledge as the conditional probability, such as the one in [16]. However, our learning model does not require any priori knowledge about the opponent and works based only on the historical offers received until t_b from the opponent. By comparing the fitted points $\hat{O}_{t_b}$ on the regression line based on each random reservation point X_i with the historical offers O_{t_b}, the conditional probability $P(O|H_i)$ is obtained. The more consistent the fitted offers are with opponent's historical offers, the higher the conditional probability $P(O|H_i)$ will be. As showed at Step 3 in Subsection 3.2.1, the difference between the regression curve and opponent's bidding sequence can be indicated by the non-linear correlation coefficient γ. Thus, we can use the value of γ as the conditional probability.

The learning approach will increase the probability of a hypothesis when the random reservation point selected in the detecting cell is most consistent with the real reservation point of the opponent. However, in some cases, it is possible that the learning may have errors. As seen in Figure 2, compared with point X_5, point X_4 has a higher non-linear correlation with the real reservation point X, but point X_4 and X are not in the same detecting cell. As a result, the hypothesis that the real reservation point X belongs to the cell where point X_4 is located has a higher probability. Nevertheless, we claim that this situation does not affect the learning effectiveness based on the following two considerations. Firstly, although in certain circumstances, using the non-linear correlation to calculate the difference between the regression line and the real bidding sequence does not necessarily reveal the real situation, the error will be eased through Bayesian learning from a probabilistic point of view. Secondly, even the error exists, the learning approach still works because we only need to find an approximate range of the reservation point, not the precise value of opponent's reservation point. In some cases, the real reservation point X might not be located in the whole detecting region, but those cells which are closer to point X will still have a higher probability compared with other cells.

Another issue that should be taken into account is the learning rate and efficiency. At the early stage of leaning, the hypotheses space can be quite large depending on the value of *DetReg* and N (recall Subsection 3.1). It is time consuming to keep all the hypotheses in the searching space. Some hypotheses can be precluded from the hypotheses space when the current time and opponent's bidding value have surpassed the detecting cell boundary. For example, for a cell $C_i = [t_i^l, t_i^h, p_i^l, p_i^h]$, if current negotiation time $t_b > t_i^h$, the hypothesis based on this cell is meaningless because the negotiation process has already proved it false.

3.3 The Adaptive Concession Strategy

Through regression analysis and Bayesian learning stated above, a more precise estimation of the opponent's reservation point is derived, represented by the renewed belief of the probability distribution of the hypothesis H_i. Now, the agent needs to take an action to give a counter offer based on this new belief, i.e. which concession strategy to take and how strong it should be in terms of a value of the concession parameter β. Our adaptive concession strategy includes two parts: the optimal concession strategy described in Subsection 3.3.1 and the combining mechanism described in Subsection 3.3.2.

3.3.1 The Optimal Concession Strategy

There are four scenarios according to different location of the random reservation point. As we believe that the agent is rational, it always strives for a highest utility of its own regardless of its opponent fully. Therefore, in each scenario, the buyer needs to adopt different concession strategies to maximize its expected utility as depicted in Figure 3. In Figure 3, point $b_0(t_0, p_0)$ is buyer's current offer at time t_0, point $b_T(T_b, RP_b)$ is the buyer's reservation offer at deadline T_b, and point $X_i(t_i^x, p_i^x)$ is the random reservation point of seller. Then the buyer needs to find another point $P(t_p, p_p)$, which is called a concession point, in its negotiation region to set the concession strategy and the value of β.

- **Scenario 1:** $(t_i^x < T_b)$ and $(p_i^x > p_0)$.
  In this scenario, the random reservation point X_i is in the buyer's negotiation region. Because the buyer agent is rational, it will always try to gain the maximal utility itself. If the buyer knows that the seller will quit the negotiation at point X_i (i.e., the deadline of the seller t_i^x is shorter than its deadline T_b), the optimal concession strategy for the buyer is to set his bidding price to p_i^x at time t_i^x. Otherwise, if the buyer gives more concession, it cannot achieve the maximal utility after finishing negotiation. On the contrary, a less concession may result in a failure of the negotiation. As illustrated in Figure 3(a), the random reservation point X_i is set to be the concession point P in this case and the dashed line crossing point X_i is the concession line of the buyer.
- **Scenario 2:** $(t_i^x >= T_b)$ and $(p_i^x >= p_0)$.
  In this scenario, random reservation point X_i is out of the buyer's negotiation region. There are two cases in this scenario according to the different regression lines of the seller. As can be seen in Figure 3(b), in the first case, regression line l_1 traverses the buyer's negotiation region while l_2 does not. In the same way of analyzing in *Scenario 1*, buyer's optimal concession line for l_1 is to pass through the intersection point of the line l_1 and the right boundary of the buyer's negotiation region. Considering that the buyer should give out its reserve price at deadline T_b, for simplicity, we let the buyer's concession line cross the concession point P_1 on the regression line one step ahead of the deadline T_b (i.e., $T_b - 1$) such that a concrete value of the concession parameter β can be computed. As for the second case, the regression curve l_2 has no intersection with the buyer's negotiation region, which means even the buyer concedes, the negotiation based on this random reservation point is doomed to fail. Nevertheless, the buyer will spare no efforts to reverse this unfavorable situation. With this aim, it will give the reserve price at next round $(t_0 + 1)$. To compute a value of β, we choose a variable $\phi_{max}(0 < \phi_{max} < 1)$ which is quite close to 1. The concession point P_2 in

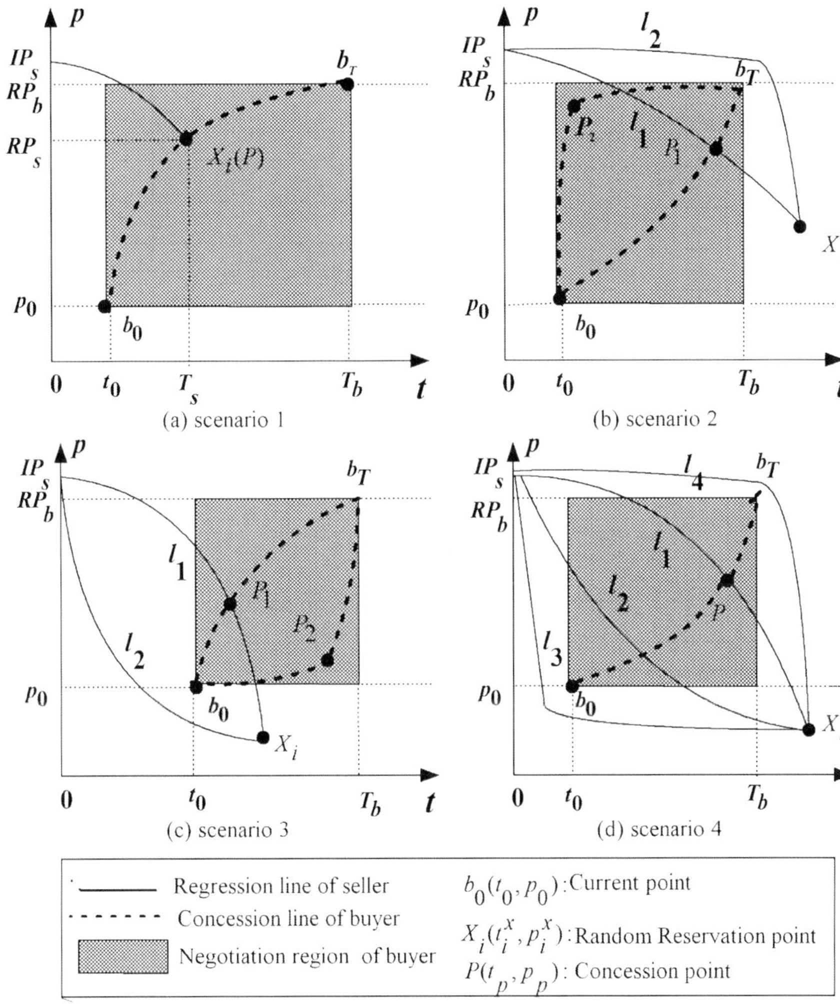

Fig. 3 Four Scenarios of Concession Strategy

this case is set to be $P_2(b_0+1, \phi_{max} \cdot RP_b)$ so as to make the price at next round close to the reserve price of RP_b and finally to give out the reserve price RP_b at the next round, i.e., its deadline.

- **Scenario 3:** $(t_i^x < T_b)$ and $(p_i^x < p_0)$.
  There are also two cases in this scenario, which can be signified by l_1 and l_2 shown in Figure 3(c). As for case 1, the optimal strategy of the buyer is to cross the intersection of l_1 and bottom line of the buyer's negotiation region. To compute a value of β, we set the concession point P_1 be the point one step earlier than the intersection point on the regression line l_1. As for case 2, the line

l_2 does not go through the buyer's negotiation region. In this case, the optimal strategy for the buyer is to keep its price unchanged until $T_b - 1$ and then gives its reserve price at the deadline. To compute the value of β, we can set the price at concession point P_2 very close to current price p_0. Similarly, a variable $\phi_{min}(0 < \phi_{min} < 1)$, which is quite close to 0, can be chosen to set the price at next round to $(1+\phi_{min}) \cdot p_0$ such that this price will keep almost the same as the current price p_0.

- **Scenario 4:** $(t_i^x >= T_b)$ and $(p_i^x <= p_0)$.
  This scenario, which is a combination of the former Scenarios 2 and 3, is the most complicated case of all. Each line of l_1, l_2, l_3 and l_4 can be analyzed in the same way as stated in the previous scenarios. In Figure 3(d), we depict the concession line based on l_1 as an example.

3.3.2 The Combining Mechanism

We have given out all possible situations of the random reservation points and the corresponding optimal concession strategies that the buyer can adopt to increase its utility as well as to avoid the failure of negotiation to its best. Because the buyer still uses the family of polynomial functions to concede, the counteroffer from point $b_0(t_0, p_0)$ can be generated by Equation 9 based on Equation 4.

$$Offer_b(t) = p_0 + (RP_b - p_0)\left(\frac{t - t_0}{T_b - t_0}\right)^{\beta} \quad (t > t_0) \tag{9}$$

Using this equation, we can guarantee that at its deadline T_b, the buyer will give the reserve price RP_b. At current time t_0, the buyer's offer is p_0 and the buyer concedes in the form of polynomial function. Then given the concession point $P(t_p, p_p)$ in its negotiation region, a new value of parameter $\hat{\beta}$ can be calculated as follows.

$$\hat{\beta} = \log_{\frac{t_p - t_0}{T_b - t_0}} \frac{p_0 - p_p}{p_0 - RP_b} \quad (t_0 < t_p < T_b) \tag{10}$$

We have calculated all the concession values $\hat{\beta}$ for each valid random reservation point in the detecting region, with a probability distribution $P(H_i) = \{p(H_1), p(H_2), ..., p(H_n)\}$ over these values derived from the regression analysis and Bayesian learning. Now comes to the problem of how to combine all the estimated value of $\hat{\beta}$ to an overall value. Let $\hat{\beta}_i$ $(i \in \{1,2,...n\})$ be the estimated concession value calculated from the concession point based on the random reservation point in cell C_i. $P(H_i)$ is the probability of the $\hat{\beta}_i$, presenting the weighting proportion of the corresponding $\hat{\beta}_i$ in all the concession values. The value of $\hat{\beta}_i$ signifies the concession degree of the agent, which can be represented by the area between the concession line and the time axis, which is called *concession area*. As can be seen from Figure 4, the concession area of $\hat{\beta}_1$ is S_1, which can be denoted by $S_{b_0\hat{\beta}_1 b_T b}$. Let S_i be the

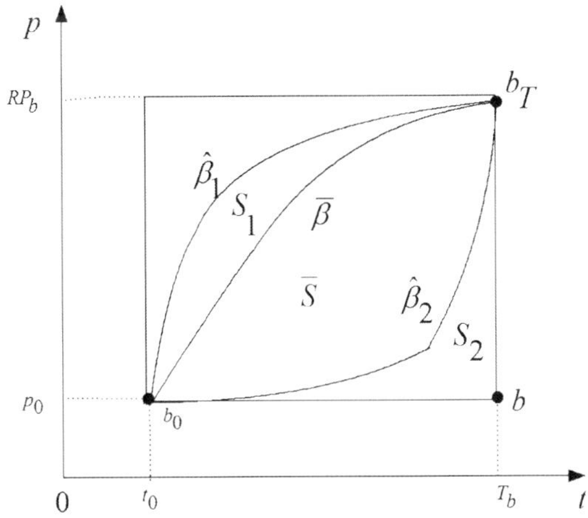

Fig. 4 Combination of the parameter β

concession area of $\hat{\beta}_i$ and let the concession area of the overall concession parameter $\bar{\beta}$ be $\bar{S}$. Based on Equation 9, we can have the following equations.

$$\bar{S} = \int_{t_0}^{T_b} [p_0 + (RP_b - P_0)(\frac{t - t_0}{T_b - t_0})^{\bar{\beta}}]d_t \tag{11}$$

$$\sum_{i=1}^{n} P(H_i)S_i = \sum_{i=1}^{n} P(H_i) \int_{t_0}^{T_b} [p_0 + (RP_b - P_0)(\frac{t - t_0}{T_b - t_0})^{\hat{\beta}}]d_t \tag{12}$$

because,

$$\bar{S} = \sum_{i=1}^{n} P(H_i)S_i \tag{13}$$

we can get the overall concession parameter $\bar{\beta}$ as follows:

$$\bar{\beta} = \frac{1}{\sum_{i=1}^{n} \frac{P(H_i)}{1+\hat{\beta}_i}} - 1 \tag{14}$$

Then the buyer can set its concession parameter as $\bar{\beta}$ to give counter offer based on Equation 9 at every step of the negotiation. Each $\hat{\beta}_i$ is changing at each step according to the randomly selected reservation point and the corresponding $P(H_i)$ is revised by Bayesian learning throughout the negotiation process. Thus the concession parameter $\bar{\beta}$ adopted by the buyer at each step is totally different, making the negotiation an adaptive process in the point view of the learning agent buyer.

4 Experiment

4.1 Experimental Setting

In the experiment, a buyer and a seller negotiate over the price ranged in-between \$0 $\sim$ \$100. In order to simplify the comparison process, we set the buyer agent's initial price to \$0 and the seller agent's initial price to \$100. The buyer's reserve price is randomly selected in-between \$50 $\sim$ \$100 and seller's reserve price is randomly selected in-between \$0 $\sim$ \$50. Such a setting ensures that the agreement zone between the two agents always exists. Our agents' deadlines are randomly selected in-between $[20, 40]$, and the concession strategies are randomly selected in-between $[0.5, 2]$. The negotiation parameter initialization is showed in Table 1.

Table 1 Negotiation Parameters initialization

Agent	IP_i	RP_i	T_i	β_i
Buyer (*i*=b)	\$0	[\$50,\$100]	[20,40]	[0.5,2]
Seller (*i*=s)	\$100	[\$0,\$50]	[20,40]	[0.5,2]

To provide a benchmark we compare our negotiation model with the general NDF model. In the general NDF model, both agents randomly initialize their negotiation parameters according to Table 1, and keep these parameters unchanged during the negotiation process. On the contrary, in our model, the buyer agent will learn how to adjust its concession strategy adaptively to reach a better negotiation outcome. To use the learning mechanism, we set $\phi_{min} = 0.01$, $\phi_{max} = 0.99$. We choose two initialization of detecting region as $DetReg1 = (0, 1.5T_b, 0, RP_b)$, $DetReg2 = (T_b, 2T_b, 0.5RP_b, 1.5RP_b)$ to test its affect on the learning result. We outline four cases according to the different numbers of detecting cells as shown in Table 2.

Table 2 Four scenarios of different detecting cell numbers

Case	N^t	N^p	N_{all}
1	4	4	16
2	8	8	64
3	16	16	256
4	20	20	400

4.2 Results and Analysis

As our model depends on the regression analysis which may yield errors as stated before, we do not expect the learning result to be completely precise. Further more, many factors affect the learning process such as the number of detecting cells, the

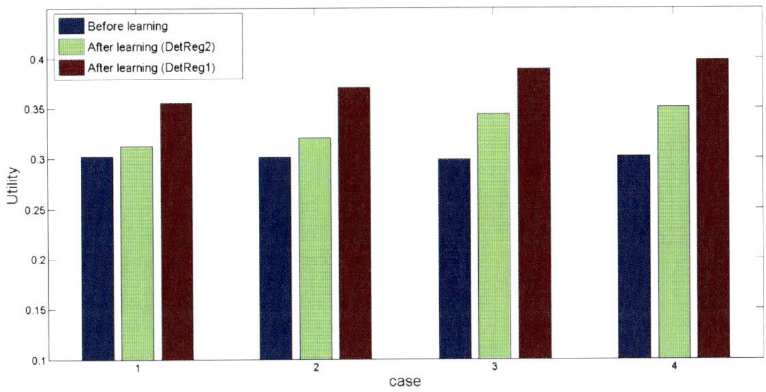

Fig. 5 The average utility in different cases before & after learning

initialization of the detecting region and different values of ϕ_{min}, ϕ_{max}. The objective of this experiment is therefore carried out to analyze the overall performance of this learning approach considering these uncertainties and potential errors.

We run 100 episodes for each case to show the generality and robustness of our model. The results of this experiment are presented in Figure 5. The x-axis indicates the four cases and the y-axis indicates the average utility of the buyer in each case. The blue bars represent the buyer's average utility gained by NDF model. The brown and green bars represent the buyer's average utility gained by using our model when the detecting region is initialized as $DetReg1$ and $DetReg2$, respectively. We can see from Figure 5, the brown and green bars are higher than the blue bars in all cases, and gradually increase as the number of the detecting cells increase. Such experimental results indicate that: (1)using our learning mechanism and the adaptive concession strategy will result in a higher utility than the static concession strategy; and (2) as the total number of detecting cells increases, the agent has a more precise estimation of the opponent's reservation point, thus can result in a higher utility. From Figure 5, we can also see that there are some difference between the learning results when the detecting region is initialized differently. More specifically, the average utilities when the detecting region is initialized as $DetReg1$ are higher than the utilities when the detecting region is initialized as $DetReg2$. This result can be explained by the fact that when the detecting region is $DetReg2$, the opponent's (seller's) real reservation point is out of the detecting region all the time. Therefore the cell with the highest probability in the detecting region (i.e., the cell that the opponent's real reservation point is most likely located in) cannot reflect the true situation, making the learning result in $DetReg2$ be inferior to that in $DetReg1$.

In order to illustrate the dynamic adaptation of the concession parameter β, we give out the whole negotiation process to show how the buyer agent changes its

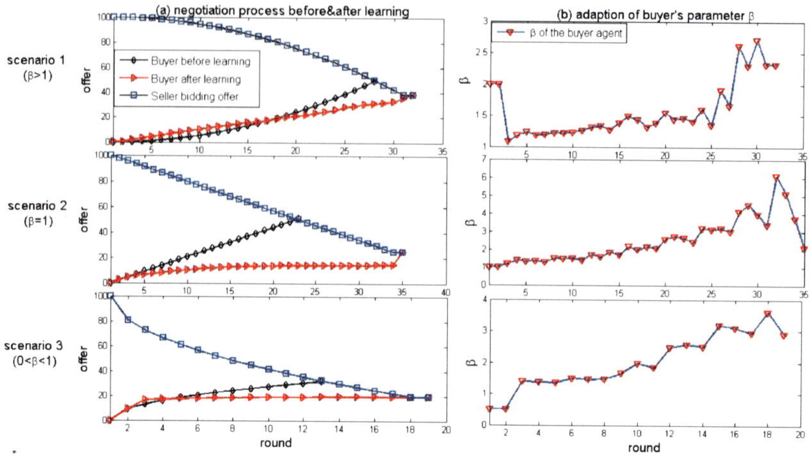

Fig. 6 The negotiation processes before & after learning

concession strategy adaptively. We select three scenarios when detecting region is *DetReg*1 with the negotiation parameters as follows:

- Scenario 1 ($1 < \beta < 2$): $RP_b = \$69.58$, $RP_s = \$11.38$, $T_b = 32$, $T_s = 36$, $\beta_b = \beta_s = 2.0$
- Scenario 2 ($\beta = 1$): $RP_b = \$81.04$, $RP_s = \$17.82$, $T_b = 35$, $T_s = 36$, $\beta_b = \beta_s = 1.0$
- Scenario 3 ($0 < \beta < 1$): $RP_b = \$50.34$, $RP_s = \$11.38$, $T_b = 30$, $T_s = 22$, $\beta_b = \beta_s = 0.5$

Figures in 6(a) give the negotiation process between both agents before & after learning. Figures in 6(b) show the adaption of the buyer's concession parameter β. In Scenario 1, the seller adopts the *Boulware* concession strategy. Before learning, the negotiation ends at \$49.53 and both agents' concession strategies keep unchanged through the negotiation process. After learning, the buyer agent adjusts its concession strategy adaptively in terms of parameter β and the agreement price is reduced to \$38.46, which is a better result than that of before learning for the buyer agent. In Scenario 2, the seller uses the *Linear* concession strategy. Before learning, the negotiation ends at \$48.57 and after learning, the buyer can have a better agreement at \$22.34. In Scenario 3, the seller uses the *Conceder* concession strategy. Before learning, the final agreement is \$33.38 and after learning this value decreases to \$20.58. According to these experimental results from the three scenarios, we can conclude that, through learning of the opponent's historical offers, the agent employing our negotiation model can effectively adapt its concession strategy so as to increase its negotiation outcome. Our negotiation model is robust when the opponent employs different concession strategies.

In this section, we illustrate the experimental results of our negotiation model and compare the results with the NDF. The experimental results indicate that our negotiation model can dynamically adapt a negotiation agent's concession strategy and significantly increase a negotiation agent's utility through the learning of the opponent's historical offers.

5 Related Work

Although incorporating learning into agent negotiation is a relatively new research topic, many approaches, models and mechanisms have been developed in recent years to solve different issues in this topic [1] [3] [6] [7] [9] [10] [11] [12] [13] [16]. We discuss several related works and compare them with our work in this paper.

Bayesian learning technique has been widely applied in negotiation for a better negotiation outcome. Zeng and Sycara were probably the first to propose a Bayesian learning based negotiation model [16]. A sequential decision making model called *Bazaar* was introduced to model beliefs of the opponent's reservation point. Our model differs from their approach in two ways. (1) *Bazaar* can only learn the reserve price of the opponent while our model can learn both opponent's price and deadline, and (2) *Bazzar* requests priori knowledge about the potential distribution of of the opponent's reserve price while our model has no this request. In [11], the authors adopted the Markov chain frameowork to model bilateral single issue negotiations among agents in dynamic environments and use Bayesian learning to enable agents to learn an optimal strategy in incomplete information settings. However, as in [16], the approach still requires the learning agent to have a prior knowledge about the conditional probability when using the Bayesian updating rule. Furthermore, this approach is designed to model the strategies of the opponent, while in our approach the agent is learning by modeling the private negotiation parameters of the opponent.

Reinforcement learning [17] is another valid technique to be integrated into negotiation [1] [14]. Soo and Hung used Q-Leaning algorithm in bilateral multi-issue negotiation [14]. However, in their work, the agent's reservation price was assumed as common knowledge. In [1], a bilateral price negotiation strategy based on Bayesian classification and Q-learning was proposed for a negotiation agent to make the best use of the opponent's negotiation history to make a decision of the opponent's classification based on Bayesian classification and then to create counter-offer efficiently by Q-learning. However, the approach in [1] is based on the classification of the opponent such that a learning agent can adjust its belief accordingly. This classification rule is set beforehand and is assumed as prior knowledge while our approach enables the agent to adapt its belief based on the learning results from the regression analysis, which are only determined by the historical offers.

Regression analysis was also employed by Ren and Zhang to predict the opponent's concession strategy by using the historical offers only [6] [7]. However, the approaches in [6] [7] did not give further advice on how to adapt agent self's concession strategy based on the learning results. Our approach moves further by contrive an efficient adaptive concession strategy based on the learning results derived from the regression analysis and Bayesian learning, making the negotiation process totally dynamic and adaptive in the view of the learning agent.

Narayanan and Jennings proposed a novel adaptive negotiation model considering the dynamism in E-commerce settings [12]. Their model manages a negotiation process as a Markov Decision Process(MDP) and uses a value iteration algorithm to acquire optimal policies to adopt different concession strategies. However, their method can only determine the adaptive action to choose a concession strategy and cannot produce a precise concession value while our model can provide constructive guidance to the agent to dynamically adaptive its behaviors including both strategies and concession values. Bzostowski and Kowalczyk [10] presented an approach for modeling behaviors of negotiators and predictive decision-making. Both their approach and our work use the similar method in term of adaptive concession strategy based only on the historical offers. However, their approach focuses more on the analysis of the differences between adjacent offers from the opponents, and will become ineffective when these differences are not significant. Our approach employs the regression analysis and will not be affected by the variance of adjacent offers.

6 Conclusion and Future Work

In this paper, we proposed an adaptive bilateral negotiation model based on Bayesian learning. This model includes a learning mechanism and an adaptive concession strategy. Through Bayesian learning, an agent's belief about the opponent's reserve price can be revised dynamically during negotiation by comparing the fitted offers derived from a regression analysis with the opponent's historical offers. The agent then proposes a countermeasure based on an adaptive concession strategy. The proposed model can enable an agent to adapt its concession strategies dynamically according to the updated probability distribution in a predicting region, thus making the negotiation process dynamic and adaptive in the view of the learning agent. The experimental results demonstrate the good performance of our model by comparison with non-learning NDF model.

There are several direction for future research. Firstly, it is imperative to extend our model to multi-issue negotiation. Considering another important factor (i.e.,weighting among the issues), our model can be further extended to make a possible win-win outcome for both agents. Secondly, we will take the non-linear utility function into account to broaden its application in practice. At last but not least, it is necessary to suit our model for more complex scenarios when the opponent is changing its deadline, reserve price, or even has a dynamic mixture of different strategies. These are all feasible and worthwhile aspects for further research. We leave them for future work.

References

1. Sun, T., Zhu, Q., Xia, Y., Cao, F.: A Bilateral Price Negotiation Strategy Based on Bayesian Classification and Q-learning. Journal of Information & Computational Science 13, 2773–2780 (2011)
2. Ng, S., Sulaiman, M., Selamat, M.: Intelligent negotiation agents in electronic commerce applications. Journal of Artificial Intelligence 2, 29–39 (2009)
3. Ng, S., Sulaiman, M., Selamat, M.: Machine learning approach in optimizing negotiation agents for E-commerce. Information Technology Journal 8, 801–810 (2009)
4. Aydogan, R., Yolum, P.: Ontology-based learning for negotiation. In: Proceedings of the 2009 IEEE/WIC/ACM International Joint Conference on Web Intelligence and Intelligent Agent Technology, vol. 02, pp. 177–184. IEEE Computer Society (2008)
5. Hindriks, K., Tykhonov, D.: Opponent Modelling in Automated Multi-Issue Negotiation Using Bayesian Learning. In: Proceedings of the 7th International Conference on Autonomous Agents and Multi Agent Systems, AAMAS 2008, pp. 331–338. IFAAMAS, Richland (2008)
6. Ren, F., Zhang, M.-J.: Predicting Partners' Behaviors in Negotiation by Using Regression Analysis. In: Zhang, Z., Siekmann, J.H. (eds.) KSEM 2007. LNCS (LNAI), vol. 4798, pp. 165–176. Springer, Heidelberg (2007)
7. Ren, F., Zhang, M.: Prediction of Partners' Behaviors in Agent Negotiation under Open and Dynamic Environments. In: 2007 IEEE/WIC/ACM International Conferences on Web Intelligence and Intelligent Agent Technology Workshops, pp. 379–382. IEEE (2007)
8. Aydogan, R., Yolum, P.: Learning consumer preferences using semantic similarity. In: Proceedings of Sixth International Joint Conference on Autonomous Agents and Multiagent Systems, AAMAS 2007, pp. 1293–1300. ACM, New York (2007)
9. Ramchurn, S.D., Sierra, C., Godo, L., Jennings, N.R.: Negotiating using rewards. Artificial Intelligence 171, 805–837 (2007)
10. Bzostowski, J., Kowalczyk, R.: Predicting partner's behavior in agent negotiation. In: Proceedings of 5th International Joint Conference on Autonomous Agents and Multiagent Systems, AAMAS 2006, pp. 355–361. IFAAMAS, Richland (2006)
11. Narayanan, V., Jennings, N.R.: Learning to Negotiate Optimally in Non-stationary Environments. In: Klusch, M., Rovatsos, M., Payne, T.R. (eds.) CIA 2006. LNCS (LNAI), vol. 4149, pp. 288–300. Springer, Heidelberg (2006)
12. Narayanan, V., Jennings, N.R.: An adaptive bilateral negotiation model for e-commerce settings. In: Seventh IEEE International Conference on E-Commerce Technology, pp. 34–41. IEEE (2005)
13. Li, J., Cao, Y.D.: Bayesian learning in bilateral multi-issue negotiation and its application in MAS-based electronic commerce. In: Proceedings of the IEEE/WIC/ACM International Conference on Intelligent Agent Technology, IAT 2004, pp. 437–440 (2004)
14. Soo, V.W., Hung, C.A.: On-line incremental learning in bilateral multi-issue negotiation. In: Proceedings of the First International Joint Conference on Autonomous Agents and Multiagent Systems: Part 1, pp. 314–315. ACM, New York (2002)
15. Faratin, P., Sierra, C., Jennings, N.R.: Negotiation decision functions for autonomous agents. Robotics and Autonomous Systems 24, 159–182 (1998)
16. Zeng, D., Sycara, K.: Bayesian learning in negotiation. International Journal of Human-Computers Studies 48, 125–141 (1998)
17. Sutton, R.S., Barto, A.G.: Reinforcement learning: An introduction. MIT Press, Cambridge (1998)

Acceptance Conditions in Automated Negotiation

Tim Baarslag, Koen Hindriks, and Catholijn Jonker

Abstract. In every negotiation with a deadline, one of the negotiating parties has to accept an offer to avoid a break off. A break off is usually an undesirable outcome for both parties, therefore it is important that a negotiator employs a proficient mechanism to decide under which conditions to accept. When designing such conditions one is faced with the acceptance dilemma: accepting the current offer may be suboptimal, as better offers may still be presented. On the other hand, accepting too late may prevent an agreement from being reached, resulting in a break off with no gain for either party. Motivated by the challenges of bilateral negotiations between automated agents and by the results and insights of the automated negotiating agents competition (ANAC), we classify and compare state-of-the-art generic acceptance conditions. We focus on *decoupled* acceptance conditions, i.e. conditions that do not depend on the bidding strategy that is used. We performed extensive experiments to compare the performance of acceptance conditions in combination with a broad range of bidding strategies and negotiation domains. Furthermore we propose new acceptance conditions and we demonstrate that they outperform the other conditions that we study. In particular, it is shown that they outperform the standard acceptance condition of comparing the current offer with the offer the agent is ready to send out. We also provide insight in to why some conditions work better than others and investigate correlations between the properties of the negotiation environment and the efficacy of acceptance conditions.

1 Introduction

Negotiation is an important process to reach trade agreements, and to form alliances or resolve conflicts. The field of negotiation originates from various disciplines

Tim Baarslag · Koen Hindriks · Catholijn Jonker
Interactive Intelligence Group
Delft University of Technology
e-mail: {T.Baarslag,K.V.Hindriks,C.M.Jonker}@tudelft.nl

T. Ito et al. (Eds.): Complex Automated Negotiations, SCI 435, pp. 95–110.
springerlink.com

including artificial intelligence, economics, social science, and game theory (e.g., [2, 15, 20]). The strategic–negotiation model has a wide range of applications, such as resource and task allocation mechanisms, conflict resolution mechanisms, and decentralized information services [15].

A number of successful negotiation strategies have already been established both in literature and in implementations [5, 6, 11, 12, 19]. And more recently, in 2010 seven new negotiation strategies were created to participate in the first automated negotiating agents competition (ANAC 2010) [3] in conjunction with the Ninth International Conference on Autonomous Agents and Multiagent Systems (AAMAS-10). During post tournament analysis of the results, it became apparent that different agent implementations use various conditions to decide when to accept an offer. In every negotiation with a deadline, one of the negotiating parties has to accept an offer to avoid a break off. Therefore, it is important for every negotiator to employ a mechanism to decide under which conditions to accept. However, designing a proper acceptance condition is a difficult task: accepting too late may result in the break off of a negotiation, while accepting too early may result in suboptimal agreements.

The importance of choosing an appropriate acceptance condition is confirmed by the results of ANAC 2010 (see Table 1). Agents with simple acceptance criteria were ranked at the bottom, while the more sophisticated time- and utility-based criteria obtained a higher score. For instance, the low ranking of *Agent Smith* was due to a mistake in the implementation of the acceptance condition [27].

Despite its importance, the theory and practice of acceptance conditions has not yet received much attention. The goal of this paper is to classify current approaches and to compare acceptance conditions in an experimental setting. Thus in this paper we will concentrate on the final part of the negotiation process: the acceptation of an offer. We focus on decoupled acceptance conditions: i.e., generic acceptance conditions that can be used in conjunction with an arbitrary bidding strategy.

Table 1 An overview of the rank of every agent in ANAC 2010 and the type of acceptance conditions that they employ

Rank	Agent	Acceptance condition
1	*Agent K*	Time and utility based
2	*Yushu*	Time and utility based
3	*Nozomi*	Time and utility based
4	*IAMhaggler*	Utility based only
5	*FSEGA*	Utility based only
6	*IAMcrazyHaggler*	Utility based only
7	*Agent Smith*	Time and utility based

Our contribution is fourfold:

i. We give an overview and provide a categorization of current decoupled acceptance conditions.
ii. We introduce a formal negotiation model that supports the use of arbitrary acceptance conditions.

iii. We compare a selection of current generic acceptance conditions and evaluate them in an experimental setting.
iv. We propose new acceptance conditions and test them against established acceptance conditions, using varying types of bidding techniques.

The remainder of this paper is organized as follows. Section 2 defines the model of negotiation that we employ and provides an overview of current acceptance conditions. In Section 3, we also consider combinations of acceptance conditions. Section 4 discusses our experimental setup and results, which demonstrate that some combinations outperform traditional acceptance conditions. Finally, Section 5 outlines our conclusions and our plans for further research on acceptance strategies.

2 Acceptance Conditions in Negotiation

This paper focuses on acceptance conditions (also called acceptance criteria) that are decoupled: i.e. generic acceptance conditions that are not tied to a specific agent implementation and hence can be used in conjunction with an arbitrary bidding strategy. We first describe a general negotiation model which fits current decoupled acceptance conditions. We have surveyed existing negotiation agents to examine the acceptance criteria that they employ. We then categorize them according to the input that they use in their decision making process.

2.1 Negotiation Model

We consider *bilateral* negotiations, i.e. a negotiation between two parties or agents A and B. The agents negotiate over *issues* that are part of a negotiation *domain*, and every issue has an associated range of alternatives or *values*. A negotiation outcome consists of a mapping of every issue to a value, and the set Ω of all possible outcomes is called the *outcome space*. The outcome space is common knowledge to the negotiating parties and stays fixed during a single negotiation session.

We further assume that both parties have certain preferences prescribed by a *preference profile* over Ω. These preferences can be modeled by means of a utility function U, which maps a possible outcome $\omega \in \Omega$ to a real-valued number in the range $[0,1]$. In contrast to the outcome space, the preference profile of the agents is private information.

Finally, the interaction between negotiating parties is regulated by a *negotiation protocol* that defines the rules of how and when proposals can be exchanged. We use the alternating-offers protocol [23] for bilateral negotiation, in which the negotiating parties exchange offers in turns.

As in [26], we assume a common global time, represented here by $\mathscr{T} = [0,1]$. We supplement the alternating-offers protocol with a deadline $t = 1$, at which moment both agent receive utility 0. This is the same setup as [8], with the exception that issues are not necessarily real-valued and both agents have the same deadline equal to $t = 1$. We represent by $x^t_{A\to B}$ the negotiation outcome proposed by agent A to agent B at time t. A *negotiation thread* (cf. [5, 26]) between two agents A and B at time $t \in \mathscr{T}$ is defined as a finite sequence

$$H^t_{A\leftrightarrow B} := \left(x^{t_1}_{p_1\to p_2}, x^{t_2}_{p_2\to p_3}, x^{t_3}_{p_3\to p_4}, \ldots, x^{t_n}_{p_n\to p_{n+1}}\right),$$

where

i. $t_k \leq t_l$ for $k \leq l$, the offers are ordered over time $\mathscr{T}$,
ii. $p_k = p_{k+2} \in \{A, B\}$ for all k, the offers are alternating between the agents,
iii. All t_i represent instances of time $\mathscr{T}$, with $t_n \leq t$,
iv. $x^{t_k}_{p_k\to p_{k+1}} \in \Omega$ for $k \in \{1, \ldots, n\}$, the agents exchange complete offers.

Additionally, the last element of $H^t_{A\leftrightarrow B}$ may be equal to one of the particles $\{Accept, End\}$. We will say a negotiation thread is *active* if this is not the case.

When agent A receives an offer $x^t_{B\to A}$ from agent B sent at time t, it has to decide at a later time $t' > t$ whether to accept the offer, or to send a counter-offer $x^{t'}_{A\to B}$. Given a negotiation thread $H^t_{A\leftrightarrow B}$ between agents A and B, we can formally express the action performed by A with an *action function* X_A:

$$X_A(t', x^t_{B\to A}) = \begin{cases} End & \text{if } t' \geq 1 \\ Accept & \text{if } \mathbf{AC}_A(t', x^{t'}_{A\to B}, H^t_{A\leftrightarrow B}) \\ x^{t'}_{A\to B} & \text{otherwise} \end{cases}$$

Note that we extend the setting of [8, 26] by introducing the *acceptance condition* $\mathbf{AC}_A$ of an agent A. This model enables us to study arbitrary decoupled acceptance conditions. $\mathbf{AC}_A$ that takes as input

$$\mathscr{I} = (t', x^{t'}_{A\to B}, H^t_{A\leftrightarrow B}),$$

the tuple containing the current time t', the offer $x^{t'}_{A\to B}$ that the agent considers as a bid (in line with the bidding strategy the agent uses), and the ongoing negotiation thread $H^t_{B\leftrightarrow A}$.

The resulting action given by the function $X_A(t', x^t_{B\to A})$ is used to extend the current negotiation thread between the two agents. If the agent does not accept the current offer, and the deadline has not been reached, it will prepare a counter-offer $x^{t'}_{A\to B}$ by using a bidding strategy or *tactic* to generate new values for the negotiable issues. Tactics can take many forms, e.g. time-dependent, resource dependent, imitative, and so on [26]. In our setup we will consider the tactics as given and try to optimize the accompanying acceptance conditions.

2.2 Acceptance Criteria

Let an active negotiation thread

$$H^t_{A\leftrightarrow B} = \left(x^{t_1}_{p_1\to p_2}, x^{t_2}_{p_2\to p_3}, \ldots, x^{t_{n-1}}_{A\to B}, x^{t_n}_{B\to A}\right),$$

be given at time $t' > t = t_n$, so that it is agent A's turn to perform an action.

As outlined in our negotiation model, the action function X_A of an agent A uses an acceptance condition $\mathbf{AC}_A(\mathscr{I})$ to decide whether to accept. In practice, most agents do not use the full negotiation thread to decide whether it is time to accept. For instance many agent implementations, such as [7, 8, 26], use the following implementation of $\mathbf{AC}_A(\mathscr{I})$:

$$\mathbf{AC}_A(t', x^{t'}_{A\to B}, H^t_{A\leftrightarrow B}) \iff U_A(x^t_{B\to A}) \geq U_A(x^{t'}_{A\to B}).$$

That is, A will accept when the utility U_A for the opponent's last offer at time t is greater than the value of the offer agent A is ready to send out at time t'. The acceptance condition above depends on the agent's upcoming offer $x^{t'}_{A\to B}$. For $\alpha, \beta \in \mathbb{R}$ this may be generalized as follows:

$$\mathbf{AC}^{\mathscr{I}}_{\text{next}}(\alpha, \beta) \overset{\text{def}}{\iff} \alpha \cdot U_A(x^t_{B\to A}) + \beta \geq U_A(x^{t'}_{A\to B}).$$

We can view α as the scale factor by which we multiply the opponent's bid, while β specifies the minimal 'utility gap' [12] that is sufficient to accept.

Analogously, we have acceptance conditions that rely on the agent's *previous* offer $x^{t_{n-1}}_{A\to B}$:

$$\mathbf{AC}^{\mathscr{I}}_{\text{prev}}(\alpha, \beta) \overset{\text{def}}{\iff} \alpha \cdot U_A(x^t_{B\to A}) + \beta \geq U_A(x^{t_{n-1}}_{A\to B}).$$

Note that this acceptance condition does not take into account the time that is left in the negotiation, nor any offers made previous to time t. Other acceptance conditions may rely on other measures, such as the remaining negotiation time or the utility of our previous offer. For example, there is a very simple acceptance criterion that only compares the opponent's offer with a constant α:

$$\mathbf{AC}^{\mathscr{I}}_{\text{const}}(\alpha) \overset{\text{def}}{\iff} U_A(x^t_{B\to A}) \geq \alpha.$$

Last but not least, instead of considering utility agents may employ a time-based condition to accept after a certain amount of time $T \in \mathscr{T}$ has passed:

$$\mathbf{AC}^{\mathscr{I}}_{\text{time}}(T) \overset{\text{def}}{\iff} t' \geq T.$$

We will omit the superscript $\mathscr{I}$ when it is clear from the context. We will use these general acceptance conditions to classify existing acceptance mechanisms in the next section.

2.3 Existing Acceptance Conditions

We give a short overview of decoupled acceptance conditions used in literature and current agent implementations. We are primarily interested in acceptance conditions that are not specifically designed for a single agent. We do not claim the list below is complete; however it serves as a good starting point to categorize current decoupled acceptance conditions. We surveyed the entire pool of agents of ANAC 2010, including *Agent K*, *Nozomi* [25], *Yushu* [1], *IAM(crazy)Haggler* [4], *FSEGA* [24] and *Agent Smith* [27]. We also examined well-known agents from literature, such as the Trade-off agent [6], the Bayesian learning agent [10], *ABMP* [12], equilibrium strategies of [7], and time dependent negotiation strategies as defined in [22], i.e. the Boulware and Conceder tactics.

Listed in Table 2 is a selection of generic acceptance conditions found.

Table 2 A selection of existing decoupled acceptance conditions found in literature and current agent implementations

AC	α	β	**Agent**
$\mathbf{AC}_{\text{prev}}(\alpha,\beta)$	1.03	0	*FSEGA*, *Bayesian Agent*
	1	0	*Agent Smith*
	1.02	0	*IAM(crazy)Haggler*
	1	0.02	*ABMP*
$\mathbf{AC}_{\text{next}}(\alpha,\beta)$	1	0	*FSEGA*, Boulware, Conceder, Trade-off, Equilibrium strategies
	1.02	0	*IAM(crazy)Haggler*
	1.03	0	*Bayesian Agent*
$\mathbf{AC}_{\text{const}}(\alpha)$	1	-	*FSEGA*
	0.9	-	*Agent Smith*
	0.88	-	*IAM(crazy)Haggler*
	T		
$\mathbf{AC}_{\text{time}}(T)$	0.92	-	*Agent Smith*

Some agents also use logical combinations of different acceptance conditions at the same time. This explains why some agents are listed multiple times in the table. For example, both *IAMHaggler* and *IAMcrazyHaggler* [4] accept precisely when

$$\mathbf{AC}_{\text{const}}(0.88) \vee \mathbf{AC}_{\text{next}}(1.02,0) \vee \mathbf{AC}_{\text{prev}}(1.02,0).$$

We will not focus on the many possible combinations of all acceptance conditions that may thus be obtained; we will study the basic acceptance conditions in isolation with varying parameters. However in addition to this we study a small selection of combinations in Section 3. We leave further combinations for future research.

As can be seen from Table 2, in our sample the most commonly used acceptance condition is $\mathbf{AC}_{\text{next}} = \mathbf{AC}_{\text{next}}(1,0)$, which is the familiar condition of accepting when the opponent's last offer is better than the planned offer of the agent. The function $\beta \mapsto \mathbf{AC}_{\text{prev}}(1,\beta)$ can be viewed as an acceptance condition that accepts when the *utility gap* [12] between the parties is smaller than β. We denote this condition by $\mathbf{AC}_{\text{gap}}(\beta)$.

3 Combined Acceptance Conditions

We define three acceptance conditions that are designed to perform well in conjunction with an arbitrary bidding strategy. This will incorporate all ideas behind the traditional acceptance conditions we have described so far. We will show in Section 4 that they work better than the majority of simple generic conditions listed in Table 2.

From a negotiation point of view, it makes sense to alter the behavior of an acceptance condition when time is running short. For example, many ANAC agents such as *Yushu*, *Nozomi* and *FSEGA* [1, 24, 25] split the negotiation into different intervals of time and apply different sub-strategies to each interval.

The basic idea behind combined acceptance conditions $\mathbf{AC}_{\text{combi}}$ is similar. In case the bidding strategy plans to propose a deal that is worse than the opponent's offer, we have reached a consensus with our opponent and we accept the offer. However, if there still exists a gap between our offer and time is short, the acceptance condition should wait for an offer that is not expected to improve in the remaining time. Thus $\mathbf{AC}_{\text{combi}}$ is designed to be a proper extension of $\mathbf{AC}_{\text{next}}$, with adaptive behavior based on recent bidding behavior near the deadline.

To define $\mathbf{AC}_{\text{combi}}$, suppose an active negotiation thread

$$H^t_{A\leftrightarrow B} = \left(x^{t_1}_{p_1\to p_2}, x^{t_2}_{p_2\to p_3}, \ldots, x^{t_{n-1}}_{A\to B}, x^{t_n}_{B\to A}\right),$$

is given at time $t' > t = t_n > \frac{1}{2}$ near the deadline, when it is agent A's turn. Note that there is $r = 1 - t'$ time remaining in the negotiation, which we will call the *remaining time window*. A good sample of what might be expected in the remaining time window consists of the bids that were exchanged during the previous time window $W = [t' - r, t'] \subseteq \mathscr{T}$ of the same size.

Let

$$H^W_{B\to A} = \left\{x^s_{B\to A} \in H^t_{A\leftrightarrow B} \mid s \in W\right\}$$

denote all bids offered by B to A in time window W. We can now formulate the average and maximum utility that was offered during the previous time window in the negotiation thread $H = H^W_{B\to A}$:

$$\text{MAX}^W = \max_{x\in H} U_A(x).$$

and

$$\mathrm{AVG}^W = \frac{1}{|H|} \sum_{x \in H} U_A(x).$$

We let $\mathbf{AC}_{\mathrm{combi}}(T,\alpha)$ accept at time t' exactly when the following holds: $\mathbf{AC}_{\mathrm{next}}$ indicates that we have to accept, *or* we have almost reached the deadline ($t' \geq T$) and the current offer suffices (i.e. better than α) given the remaining time:

$$\mathbf{AC}_{\mathrm{combi}}(T,\alpha) \overset{\mathrm{def}}{\iff} \mathbf{AC}_{\mathrm{next}} \vee \mathbf{AC}_{\mathrm{time}}(T) \wedge (U_A(x^t_{B \to A}) \geq \alpha).$$

Note that we have defined $\mathbf{AC}_{\mathrm{combi}}(T,\alpha)$ in such a way that it splits the negotiation time into two phases: $[0,T)$ and $[T,1]$, with different behavior in both cases.

We will consider three different combined acceptance conditions:

i. $\mathbf{AC}_{\mathrm{combi}}(T,\mathrm{MAX}^W)$: the current offer is good enough when it is better than all offers seen in the previous time window W,
ii. $\mathbf{AC}_{\mathrm{combi}}(T,\mathrm{AVG}^W)$: the offer is better than the average utility of offers during the previous time window W,
iii. $\mathbf{AC}_{\mathrm{combi}}(T,\mathrm{MAX}^{\mathcal{T}})$: the offer should be better than any bid seen before.

4 Experiments

In order to experimentally test the efficacy of an acceptance condition, we considered a negotiation setup with the following characteristics. We equipped a set of agents (as defined later) with an acceptance condition, and measured the result against other agents in the following way. Suppose agent A is equipped with acceptance condition $\mathbf{AC}_A$ and negotiates with agent B. The two parties may reach a certain outcome $\omega \in \Omega$, for which A receives the associated utility $U_A(\omega)$. The score for A is averaged over all trials on various domains (see Section 4.1.2), alternating between the two preference profiles defined on that domain. E.g., on the negotiation scenario between England and Zimbabwe, A will play both as England and as Zimbabwe against all others.

For our experimental setup we employed GENIUS (General Environment for Negotiation with Intelligent multi-purpose Usage Simulation) [16]. This environment, which is also used in ANAC, helps to facilitate the design and evaluation of automated negotiators' strategies. It can be used to simulate tournaments between negotiating agents in various negotiation scenarios, such as the setup described in this section. It supports the alternating offer protocol with a real-time deadline as outlined in our negotiation model. The default negotiation time in GENIUS and in the setup of ANAC is 3 minutes per negotiation session; therefore we use the same value in our experiments.

4.1 Detailed Experimental Setup

4.1.1 Agents

We use the negotiation tactics that were submitted to The Automated Negotiating Agents Competition (ANAC 2010) [3]. ANAC is a negotiation competition aiming to facilitate and coordinate the research into proficient negotiation strategies for bilateral multi-issue negotiation, similar to what the Trading Agent Competition (TAC) has achieved for the trading agent problem [28].

The seven agents that participated in ANAC 2010 have been implemented by various international research groups of negotiation experts. We used these strategies in our experiments as they are representative of the current state-of-the-art in automated negotiation. Firstly, we removed the built-in acceptance mechanism from this representative group of agents; this left us with its pure bidding tactics. As outlined in our negotiation model, this procedure allowed us to test arbitrary acceptance conditions in tandem with any ANAC tactic.

We aimed to tune our acceptance conditions to the top performing ANAC 2010 agents. Therefore we have selected the top 3 of ANAC agents that were submitted by different research groups, namely *Agent K*, *Yushu* and *IAMhaggler* (we omitted *Nozomi* as the designing group also implemented *Agent K*, cf. Table 1). For the set of opponents, we selected all agents from ANAC 2010, for the acceptance conditions should be tested against a wide array of strategies. The opponents also had their built-in acceptance conditions removed (and hence were not able to accept), so that differences in results would depend entirely on the acceptance condition under consideration. To test the efficacy of an acceptance condition, we equipped the top 3 tactics with this condition and compared the average utility obtained by the three agents when negotiating against their opponents.

4.1.2 Domains

The specifics of a negotiation domain can be of great influence on the negotiation outcome [9]. Acceptance conditions have to be assessed on negotiation domains of different size and complexity. Negotiation results also depend on the *opposition* of the parties' preferences. The notion of weak and strong opposition can be formally defined [13]. Strong opposition is typical of competitive domains, when a gain for one party can be achieved only at a loss for the other party. Conversely, weak opposition means that both parties achieve either losses or gains simultaneously.

With this in mind, we aimed for two domains (with two preference profiles each) with a good spread of negotiation characteristics. We picked two domains from the three that were used in ANAC 2010 (cf. [3]). Some agents participating in ANAC 2010 did not scale well and could not deal with a large bid space. We omitted the Travel domain as the agents had too many difficulties with it to make it a reliable testing domain.

Our first scenario is taken from [14], which describes a buyer–seller business negotiation. It involves representatives of two companies: Itex Manufacturing, a producer of bicycle components and Cypress Cycles, a builder of bicycles. There are four issues that both sides have to discuss: the price of the components, delivery times, payment arrangements and terms for the return of possibly defective parts. The opposition between the parties is strong in this domain, as the manufacturer and consumer have naturally opposing requirements. Altogether, there are 180 potential offers that contain all combinations of values for the four issues.

The second domain taken from [17, 18] involves a case where England and Zimbabwe negotiate an agreement on tobacco control. The leaders of both countries must reach an agreement on five issues. England and Zimbabwe have contradictory preferences for the first two issues, but the other issues have options that are jointly preferred by both sides. The domain has a total of 576 possible agreements.

To compensate for any utility differences in the preference profiles, the agents play both sides of every scenario.

Table 3 The four preference profiles used in experiments

	Itex–Cyp	**Zim–Eng**
Size	180	576
Opposition	Strong	Medium

4.1.3 Acceptance Conditions

For each acceptance condition we tested all $3 \times 7 = 21$ pairings of agents, playing with each of the 4 different preference profiles. We ran every experiment twice, so that altogether each acceptance condition was tested 168 times. We selected the following acceptance conditions for experimental testing. The different values of parameters will be discussed in the section below.

- $\mathbf{AC}_{\text{next}}(\alpha, 0)$ and $\mathbf{AC}_{\text{prev}}(\alpha, 0)$ for $\alpha \in \{1, 1.02\}$,
- $\mathbf{AC}_{\text{gap}}(\alpha)$ for $\alpha \in \{0.02, 0.05, 0.1, 0.2\}$,
- $\mathbf{AC}_{\text{const}}(\alpha)$ for $\alpha \in \{0.8, 0.9\}$,
- $\mathbf{AC}_{\text{time}}(T)$, and the combined acceptance conditions $\mathbf{AC}_{\text{combi}}(T, \text{MAX}^W)$, $\mathbf{AC}_{\text{combi}}(T, \text{AVG}^W)$ and $\mathbf{AC}_{\text{combi}}(T, \text{MAX}^{\mathcal{T}})$, where W is the previous time window with respect to the current time t', and $T = 0.99$ (this particular value of T is discussed below).

Additionally, we ran the experiments with agents having their built-in acceptance mechanism in place. That is, we also tested the original agents' *coupled* acceptance mechanism. As we cannot for example, equip *Agent K* with the coupled acceptance condition of *Yushu*, we tested the built-in mechanism by having each agent employ its own mechanism.

4.2 Hypotheses and Experimental Results

The experiments considered here are designed to discuss the main properties and drawbacks of the acceptance conditions listed above. We formulate several hypotheses with respect to the acceptance conditions we have discussed.

Our hypothesis about $\mathbf{AC}_{\text{const}}(\alpha)$ is the following:

Hypothesis 1. For α close to one, $\mathbf{AC}_{\text{const}}(\alpha)$ performs worse than all other conditions.

To evaluate this hypothesis and others below, we have carried out a large number of experiments. The results are summarized in Table 4 of the appendix. The table shows the average utility obtained by the agents when equipped with several acceptance conditions. The "average utility of agreements" column represents the average utility obtained by the agent given the fact that they have reached an agreement. When they do not reach an agreement (due to the deadline), they get zero utility. Thus the following holds:

(The acceptance dilemma)

$$\text{Total average utility} = \text{Agreement percentage} \times \text{Average utility of agreements.}$$

This formula captures the essence of the acceptance dilemma: accepting bad to mediocre offers yields more agreements of relatively low utility. While accepting only the best offers produces less agreements, but of higher utility.

Now consider $\mathbf{AC}_{\text{const}}(0.9)$ and $\mathbf{AC}_{\text{const}}(0.8)$. When it reaches an agreement, it receives a very high utility (at least 0.9 or 0.8 respectively), but this happens so infrequently (resp. 26% and 38% of all negotiations), that it is ranked at the bottom when we consider total average utility.

We can conclude that our hypothesis is confirmed: in isolation, $\mathbf{AC}_{\text{const}}(\alpha)$ is not very advantageous to use. The main reason is that the choice of the constant α is highly domain-dependent. A very cooperative domain may have multiple win–win outcomes with utilities above α. $\mathbf{AC}_{\text{const}}(\alpha)$ would then accept an offer which is *relatively* bad, i.e. it could have done much better. On the other hand, in highly competitive domains, it may simply 'ask for too much' and may rarely obtain an agreement. Its value lies mostly in using it in combination with other acceptance conditions such as $\mathbf{AC}_{\text{next}}$. It can then benefit the agent by accepting an unexpectedly good offer or a mistake by the opponent.

As we discussed earlier in Section 2.3, the acceptance conditions $\mathbf{AC}_{\text{prev}}(\alpha, 0)$ and $\mathbf{AC}_{\text{next}}(\alpha, 0)$ are standard in literature for $\alpha \in \{1, 1.02\}$. Many agents tend to use these acceptance conditions, as they are well-known and easy to implement. We have formed the following hypothesis:

Hypothesis 2. $\mathbf{AC}_{\text{next}}(\alpha,0)$ will outperform $\mathbf{AC}_{\text{prev}}(\alpha,0)$ for $\alpha \in \{1, 1.02\}$. However, both conditions will perform worse than conditions that take the remaining time into account.

To test this hypothesis, we consult Table 4 where we have considered the two values for α. The first observation is that $\mathbf{AC}_{\text{prev}}(\alpha,0)$ and $\mathbf{AC}_{\text{next}}(\alpha,0)$ already perform much better than $\mathbf{AC}_{\text{const}}$. The higher value for α yields a better result and $\mathbf{AC}_{\text{next}}(\alpha,0)$ does indeed outperform $\mathbf{AC}_{\text{prev}}(\alpha,0)$. It makes sense that comparing the opponent's offer to our upcoming offer is more beneficial than comparing it to our previous offer, as $\mathbf{AC}_{\text{next}}$ is always 'one step ahead' of $\mathbf{AC}_{\text{prev}}$. However, all time-dependent acceptance conditions outperform both of them, even for $\alpha = 1.02$. This also settles the second part of the hypothesis. The reason for this bad performance is that many bidding strategies focus on the 'negotiation dance' [21]. That is, modeling the opponent, trying to make equal concessions and so on. When a strategy does not explicitly take time considerations into account when making an offer, this poses a problem for the two standard acceptance conditions: they rely completely on the bidding strategy to concede to the opponent before the deadline occurs. When the agent or the opponent does not concede enough near the deadline, the standard conditions lead to poor performance.

Our third hypothesis with respect to the time-dependent condition is as follows:

Hypothesis 3. $\mathbf{AC}_{\text{time}}(T)$ always reaches an agreement, but of relatively low utility.

To evaluate this hypothesis we needed to provide a concrete value for the experimental variable T. We have set $T = 0.99$ for every acceptance condition depending on T. As we have found during preliminary experiments, this value is sufficiently close to the deadline, while it still allows enough time to reach a win-win outcome. From observing the acceptance probability of $\mathbf{AC}_{\text{time}}(0.99)$ in the experimental results, we see that in 1 out of 168 negotiations ($\approx 1\%$) this criterion did not reach an agreement due to agent crashes and protocol errors, in which case both agents received utility zero. But except for these particular events, $\mathbf{AC}_{\text{time}}(T)$ will always reach an agreement, therefore we consider this part of the hypothesis confirmed.

$\mathbf{AC}_{\text{time}}(T)$, with T close to 1 is a sensible criterion to avoid a break off at all cost. It is rational to prefer any outcome over a break off of zero utility. However, the resulting deal can be anything. As we can see from the table, this is the reverse situation of $\mathbf{AC}_{\text{gap}}$: $\mathbf{AC}_{\text{time}}(T)$ yields the lowest agreement score (0.622) of all conditions. This can be explained by the acceptance dilemma: by accepting any offer near the deadline, it reaches more agreements but of relatively low utility. Still the overall score is almost the same (0.618) and thus reasonable. It is interesting to note that $\mathbf{AC}_{\text{time}}(T)$ outperforms both $\mathbf{AC}_{\text{prev}}$ and $\mathbf{AC}_{\text{next}}$ in average overall score.

This insight led us to believe that more consideration has to be given to the remaining time when deciding to accept an offer. The combined acceptance conditions evaluated in the next chapter expand upon this idea to get better deals near the deadline.

4.2.1 Evaluating $\mathbf{AC}_{\text{combi}}(T,\alpha)$

When evaluating $\mathbf{AC}_{\text{combi}}(T,\alpha)$, we expect the following characteristics. $\mathbf{AC}_{\text{combi}}(T,\alpha)$ is an extension of $\mathbf{AC}_{\text{next}}$ in the sense that it will accept under broader circumstances. It alleviates some of the mentioned drawbacks of $\mathbf{AC}_{\text{next}}$ by also accepting when the utility gap between the parties is positive. Also note, that in addition to the parameters that current acceptance conditions use, such as my previous bid $x^{t_{n-1}}_{A\to B}$, my next bid $x^{t'}_{A\to B}$, the remaining time, and the opponent's bid $x^{t}_{B\to A}$, this condition employs the entire bidding history $H^{t}_{A\leftrightarrow B}$ to compute the acceptability of an offer. Therefore we expect better results than with $\mathbf{AC}_{\text{next}}$, with more agreements, and when it agrees, we expect a better deal than by using $\mathbf{AC}_{\text{time}}(T)$.

We capture this last statement in our final hypothesis:

Hypothesis 4. The combination $\mathbf{AC}_{\text{combi}}(T,\alpha)$ outperform other acceptance conditions, such as $\mathbf{AC}_{\text{time}}(T)$ and $\mathbf{AC}_{\text{next}}$ primarily by getting deals of higher utility.

As is evident from the experimental results, $\mathbf{AC}_{\text{combi}}(\text{MAX}^{W})$ as well as $\mathbf{AC}_{\text{combi}}(\text{AVG}^{W})$ dominate the other acceptance conditions. They even perform 7% better than the built-in mechanisms of the agent, and 18% better than $\mathbf{AC}_{\text{next}}$. Similar to $\mathbf{AC}_{\text{time}}$, both conditions still get a deal almost every time, but with a higher utility. However, the average utility of an agreement is not the highest: the $\mathbf{AC}_{\text{gap}}$ conditions and the built-in mechanisms get better agreements. But again, we can observe that their agreement rate is also lower, resulting in a higher overall score for the combined criteria.

Aiming for the highest utility that has been offered so far (i.e.: using $\mathbf{AC}_{\text{combi}}(\text{MAX}^{\mathscr{T}})$) is a less successful criterion, mostly due to a big decrease in agreements. The higher utility that is obtained with this condition does not compensate for the loss of utility that is caused by a break off.

4.3 Related Work

All existing negotiation agent implementations deal with the problem of when to accept. In many cases, the agent accepts a proposal when the value of the offered contract is higher than the offer it is ready to send out at that moment in time. Examples include the time dependent negotiation strategies defined in [22] (e.g. the Boulware and Conceder tactics). The same principle is used in the equilibrium strategies of [7] and for the Trade-off agent [6], although in this setting, the deadline can be different for both agents. In our work, we consider strategies that do not always reach an agreement, and hence we have concentrated on acceptance conditions that yield better results in such cases.

Of all ANAC 2010 participants, we shortly discuss *Agent K* [25] as it employs the most sophisticated method to decide when to accept. Its acceptance mechanism is based on the mean and variance of all received offers. It then tries to determine the best offer it might receive in the future and sets its proposal target accordingly.

In contrast to our approach, this mechanism is not fully decoupled from the bidding strategy as it directly influences its bid target. Furthermore, it does not restrict its scope to the remaining or previous time window. Finally, we note that *Agent K* performs better in our experimental setup (cf. Table 4) when equipped with our combined acceptance conditions than with its built-in mechanism.

Although we do not focus on negotiation tactics and convergence results, our negotiation model builds upon the model of [26]. However, in this model, the action function of an agent only takes into account the offer it is ready to send out at that moment in time. Moreover, the focus of the paper is not on comparing acceptance conditions as only one specific instance is studied. We take a more general approach in which the agent utilizes a generic acceptance mechanism, in which the current time and the entire bidding history is considered.

5 Conclusion and Future Work

In this paper, we aimed to classify current approaches to generic acceptance conditions and to compare a selection of acceptance conditions in an experimental setting. We presented the challenges and proposed new solutions for accepting offers in current state-of-the-art automated negotiations. The focus of this paper is on decoupled acceptance conditions, i.e. general conditions that do not depend on a particular bidding strategy.

Designing an effective acceptance condition is challenging because of the acceptance dilemma: better offers may arrive in the future, but waiting for too long can result in a break off of the negotiation, which is undesirable for both parties.

We have seen that the standard acceptance criterion $\mathbf{AC}_{\text{next}}$ is often used by negotiating agents. From our results, it is apparent that $\mathbf{AC}_{\text{next}}$ does not always yield optimal agreements. We established that it performs worse than more sophisticated acceptance conditions.

In addition to classifying and comparing existing acceptance conditions, we have devised three new acceptance conditions by combining existing ones. This included two acceptance conditions that estimate whether a better offer might occur in the future based on recent bidding behavior. These conditions obtained the highest utility in our experiments and hence performed better than the other conditions we have investigated.

A suggestion for future research would be to explore the many possible combinations of acceptance conditions that may be obtained using conjunction and disjunction (and possibly negation). Some agents already use a logical combination of different acceptance conditions at the same time. For example, the *IAM(crazy)Haggler* agents accept when

$$\mathbf{AC}_{\text{const}}(0.88) \vee \mathbf{AC}_{\text{next}}(1.02,0) \vee \mathbf{AC}_{\text{prev}}(1.02,0).$$

A suitable combination of acceptance conditions could provide a considerable improvement over current acceptance conditions.

Secondly, we plan to test acceptance conditions with more agents and on larger domains, using the resources that will be available after the upcoming ANAC 2011 event.

Finally, we did not consider negotiation domains with discount factors, which devaluate utility with the passing of time. Adding discount factors will require new acceptance conditions that give more consideration to the negotiation timeline. We plan to examine such extensions in future work.

Acknowledgements. This research is supported by the Dutch Technology Foundation STW, applied science division of NWO and the Technology Program of the Ministry of Economic Affairs. It is part of the Pocket Negotiator project with grant number VICI-project 08075.

References

1. An, B., Lesser, V.: Yushu: A Heuristic-Based Agent for Automated Negotiating Competition. In: Ito, T., Zhang, M., Robu, V., Fatima, S., Matsuo, T. (eds.) New Trends in Agent-Based Complex Automated Negotiations. SCI, vol. 383, pp. 145–149. Springer, Heidelberg (2012)
2. Aumann, R., Hart, S. (eds.): Handbook of Game Theory with Economic Applications, vol. 1. Elsevier (March 1992)
3. Baarslag, T., Hindriks, K., Jonker, C., Kraus, S., Lin, R.: The First Automated Negotiating Agents Competition (ANAC 2010). In: Ito, T., Zhang, M., Robu, V., Fatima, S., Matsuo, T. (eds.) New Trends in Agent-Based Complex Automated Negotiations. SCI, vol. 383, pp. 113–135. Springer, Heidelberg (2012)
4. Williams, C.R., Robu, V., Gerding, E.H., Jennings, N.R.: IAMhaggler: A Negotiation Agent for Complex Environments. In: Ito, T., Zhang, M., Robu, V., Fatima, S., Matsuo, T. (eds.) New Trends in Agent-Based Complex Automated Negotiations. SCI, vol. 383, pp. 151–158. Springer, Heidelberg (2012)
5. Faratin, P., Sierra, C., Jennings, N.R.: Negotiation decision functions for autonomous agents. Int. Journal of Robotics and Autonomous Systems 24(3-4), 159–182 (1998)
6. Faratin, P., Sierra, C., Jennings, N.R.: Using similarity criteria to make negotiation tradeoffs. Journal of Artificial Intelligence 142(2), 205–237 (2003)
7. Fatima, S.S., Wooldridge, M., Jennings, N.R.: Multi-issue negotiation under time constraints. In: AAMAS 2002: Proceedings of the First International Joint Conference on Autonomous Agents and Multiagent Systems, pp. 143–150. ACM, New York (2002)
8. Fatima, S.S., Wooldridge, M., Jennings, N.R.: Optimal Negotiation Strategies for Agents with Incomplete Information. In: Meyer, J.-J.C., Tambe, M. (eds.) ATAL 2001. LNCS (LNAI), vol. 2333, pp. 377–392. Springer, Heidelberg (2002)
9. Hindriks, K.V., Tykhonov, D.: Towards a Quality Assessment Method for Learning Preference Profiles in Negotiation. In: Ketter, W., La Poutré, H., Sadeh, N., Shehory, O., Walsh, W. (eds.) AMEC 2008. LNBIP, vol. 44, pp. 46–59. Springer, Heidelberg (2010)
10. Hindriks, K.V., Tykhonov, D.: Opponent modelling in automated multi-issue negotiation using bayesian learning (2008)
11. Ito, T., Hattori, H., Klein, M.: Multi-issue negotiation protocol for agents: Exploring nonlinear utility spaces (2007)
12. Jonker, C., Robu, V., Treur, J.: An agent architecture for multi-attribute negotiation using incomplete preference information. Autonomous Agents and Multi-Agent Systems 15, 221–252 (2007), doi:10.1007/s10458-006-9009-y
13. Kersten, G., Noronha, S.: Rational agents, contract curves, and inefficient compromises report. Working papers, International Institute for Applied Systems Analysis (1997)

14. Kersten, G.E., Zhang, G.: Mining inspire data for the determinants of successful internet negotiations. InterNeg Research Papers INR 04/01 Central European Journal of Operational Research (2003)
15. Kraus, S.: Strategic Negotiation in Multiagent Environments. MIT Press (October 2001)
16. Lin, R., Kraus, S., Baarslag, T., Tykhonov, D., Hindriks, K., Jonker, C.M.: Genius: An integrated environment for supporting the design of generic automated negotiators. Computational Intelligence (2012)
17. Lin, R., Kraus, S., Tykhonov, D., Hindriks, K., Jonker, C.M.: Supporting the Design of General Automated Negotiators. In: Ito, T., Zhang, M., Robu, V., Fatima, S., Matsuo, T., Yamaki, H. (eds.) Innovations in Agent-Based Complex Automated Negotiations. SCI, vol. 319, pp. 69–87. Springer, Heidelberg (2010)
18. Lin, R., Kraus, S., Wilkenfeld, J., Barry, J.: Negotiating with bounded rational agents in environments with incomplete information using an automated agent. Artificial Intelligence 172(6-7), 823–851 (2008)
19. Lin, R., Oshrat, Y., Kraus, S.: Investigating the benefits of automated negotiations in enhancing people's negotiation skills. In: AAMAS 2009: Proceedings of The 8th International Conference on Autonomous Agents and Multiagent Systems, pp. 345–352 (2009)
20. Osborne, M.J., Rubinstein, A.: Bargaining and Markets (Economic Theory, Econometrics, and Mathematical Economics). Academic Press (April 1990)
21. Raiffa, H.: The Art and Science of Negotiation. Harvard University Press (1982)
22. Ros, R., Sierra, C.: A negotiation meta strategy combining trade-off and concession moves. Autonomous Agents and Multi-Agent Systems, 163–181 (2006)
23. Rubinstein, A.: Perfect equilibrium in a bargaining model. Econometrica 50(1), 97–109 (1982)
24. Dan Şerban, L., Silaghi, G.C., Litan, C.M.: AgentFSEGA: Time Constrained Reasoning Model for Bilateral Multi-issue Negotiations. In: Ito, T., Zhang, M., Robu, V., Fatima, S., Matsuo, T. (eds.) New Trends in Agent-Based Complex Automated Negotiations. SCI, vol. 383, pp. 159–165. Springer, Heidelberg (2012)
25. Kawaguchi, S., Fujita, K., Ito, T.: AgentK: Compromising Strategy Based on Estimated Maximum Utility for Automated Negotiating Agents. In: Ito, T., Zhang, M., Robu, V., Fatima, S., Matsuo, T. (eds.) New Trends in Agent-Based Complex Automated Negotiations. SCI, vol. 383, pp. 137–144. Springer, Heidelberg (2012)
26. Sierra, C., Faratin, P., Jennings, N.: A Service-Oriented Negotiation Model Between Autonomous Agents. In: Boman, M., Van de Velde, W. (eds.) MAAMAW 1997. LNCS, vol. 1237, pp. 17–35. Springer, Heidelberg (1997)
27. van Galen Last, N.: Agent Smith: Opponent Model Estimation in Bilateral Multi-issue Negotiation. In: Ito, T., Zhang, M., Robu, V., Fatima, S., Matsuo, T. (eds.) New Trends in Agent-Based Complex Automated Negotiations. SCI, vol. 383, pp. 167–174. Springer, Heidelberg (2012)
28. Wellman, M.P., Wurman, P.R., O'Malley, K., Bangera, R., de Lin, S., Reeves, D., Walsh, W.E.: Designing the market game for a trading agent competition. IEEE Internet Computing 5(2), 43–51 (2001)

Appendix A

Table 4. Utility scores of agents equipped with different acceptance conditions

Acceptance Condition	*Agent K*	*IAMhaggler*	*Yushu*	Agreement %	Average utility of agreements	Total avg
$AC_{combi}(MAX^{W})$	0.691	0.639	0.695	99%	0.679	0.675
$AC_{combi}(AVG^{W})$	0.684	0.634	0.691	99%	0.678	0.670
$AC_{gap}(0.1)$	0.636	0.562	0.693	83%	0.761	0.630
Built-in mechanism	0.641	0.547	0.692	82%	0.768	0.627
$AC_{combi}(MAX^{\mathcal{T}})$	0.691	0.577	0.596	89%	0.696	0.621
$AC_{time}(0.99)$	0.612	0.580	0.663	99%	0.622	0.618
$AC_{gap}(0.2)$	0.626	0.579	0.650	86%	0.721	0.618
$AC_{gap}(0.05)$	0.629	0.550	0.672	78%	0.791	0.617
$AC_{next}(1.02,0)$	0.616	0.517	0.696	77%	0.788	0.610
$AC_{gap}(0.02)$	0.618	0.491	0.638	73%	0.802	0.582
$AC_{prev}(1.02,0)$	0.618	0.491	0.629	72%	0.805	0.579
$AC_{next}(1,0)$	0.586	0.517	0.597	72%	0.787	0.567
$AC_{prev}(1,0)$	0.588	0.491	0.589	69%	0.805	0.556
$AC_{const}(0.8)$	0.286	0.374	0.313	38%	0.851	0.324
$AC_{const}(0.9)$	0.215	0.272	0.231	26%	0.935	0.239

Heuristic-Based Approaches for CP-Nets in Negotiation

Reyhan Aydoğan, Tim Baarslag, Koen V. Hindriks,
Catholijn M. Jonker, and Pınar Yolum

Abstract. CP-Nets have proven to be an effective representation for capturing preferences. However, their use in multiagent negotiation is not straightforward. The main reason for this is that CP-Nets capture partial ordering of preferences, whereas negotiating agents are required to compare any two outcomes based on the request and offers. This makes it necessary for agents to generate total orders from their CP-Nets. We have previously proposed a heuristic to generate total orders from a given CP-Net. This paper proposes another heuristic based on Borda count, applies it in negotiation, and compares its performance with the previous heuristic.

1 Introduction

Modeling users' preferences is an inevitable part of automated negotiation tools. While representing the user's preferences, there are several issues to be taken into account. One, outcome space grows exponentially with the number of attributes and their possible values. It may be infeasible to ask a user to rank all outcomes when the outcome space is very large. Two, the user may have difficulty in assessing her preferences in a quantitative way [5]. Representing someone's preferences with numerical values is an arduous task for a human. Three, it is difficult to find a mathematical model for representing preferences in which there are preferential dependencies between attributes. Therefore, it is more effective and intuitive to use a qualitative preference model.

Reyhan Aydoğan · Tim Baarslag · Koen V. Hindriks · Catholijn M. Jonker
Interactive Intelligence Group, Delft University of Technology, Delft, The Netherlands
e-mail: {R.Aydogan,T.Baarslag,K.V.Hindriks}@tudelft.nl
C.M.Jonker@tudelft.nl

Pınar Yolum
Department of Computer Engineering, Boğaziçi University, Bebek, 34342, Istanbul, Turkey
e-mail: pinar.yolum@boun.edu.tr

T. Ito et al. (Eds.): Complex Automated Negotiations, SCI 435, pp. 113–123.
springerlink.com

Although it is desired for users to express their preferences qualitatively, most of the current negotiation strategies [2, 6, 7, 8, 9] work with quantitative preferences. Hence, to use qualitative preferences in negotiation, it is necessary to estimate quantitative preferences from qualitative preferences. Accordingly, this paper is about estimation of quantitative preferences from qualitative preferences. That is, we propose heuristics to allow agents to have a qualitative preference model, while their negotiation strategy employs quantitative information. In order to do so, we start from a qualitative preference representation, namely CP-Nets. CP-Nets allow representation of conditional preferences and tolerate partial ordering. We extend the GENIUS negotiation framework [11] to allow elicitation of acyclic CP-Net preferences. Then, we apply our heuristics to generate utility-based information from the given CP-Net.

We compare the performance of agents when they apply heuristics on their users' qualitative preferences and negotiate with estimated utilities versus when they have their users' real total preference orderings and negotiate with real utilities. To accomplish this, users were asked to create their preference profiles both quantitatively (UCP-Nets) and qualitatively (CP-Nets), using the GENIUS interface for an apartment renting domain. The given UCP-Nets serve as ground truth. The agents apply heuristics on the given CP-Net and then negotiate with the resulting estimated utilities. Each negotiation outcome is evaluated based on the given UCP-Net, which is not only consistent with the CP-Net but also provides a total ordering of outcomes.

The rest of this paper is organized as follows: Section 2 gives an introduction on CP-Nets and UCP-Nets. Section 3 explains the heuristics that we propose to be used with CP-Nets. Section 4 explains our experimental setup, metrics, and results. Finally, Section 5 discusses our work.

2 Background: CP-Nets and UCP-Nets

Conditional preference networks (CP-nets) is a graphical model for representing qualitative preferences in a compact way [5]. In CP-nets, each node represents an attribute and each edge denotes preferential dependency between nodes. If there is an edge from X to Y, X is called "parent node" and Y is called "child node". The preference on child nodes depends on their parent nodes' values. To express conditional preferences, each node is associated with a conditional preference table (CPT), which represents a total order on possible values of that node with respect to its parents' values.

Consider apartment renting domain in Example 1 and a CP-NET expressing that its user's preference on parking area depends on neighborhood. CPT for *Parking Area* shows that the user prefers an apartment having a parking area when the neighborhood is either *Kadikoy* or *Kartal*. However, she prefers an apartment not having a parking area when it is at *Etiler*. In CP-nets, each preference statement is interpreted under "everything else being equal" interpretation. The statement, "*Etiler* is preferred over *Kartal*", means that if all other attributes such as price and parking area are the same, an apartment at *Etiler* is preferred over an apartment at *Kartal*.

Example 1. For simplicity, we have only three attributes in our apartment renting domain: *Price*, *Neighborhood* and *Parking Area*. There are three neighborhoods: *Etiler*, *Kadikoy* and *Kartal* whereas the valid values for the price are categorized as *High*, *Medium* and *Low*. A parking area may exist or not. Thus, the domain for parking area has two values: *Yes* and *No*.

We need to check whether there exists an *improving flip* sequence from one outcome to another (and vice versa) to answer whether an outcome would be preferred over another. An improving flip is changing the value of a single attribute with a more desired value by using CPT for the attribute. If there are not any improving flip sequences between two outcomes, we cannot compare these two outcomes. Thus, the inability of comparing some outcomes is the challenge of using CP-Nets in negotiation.

Boutilier *et al.* propose UCP-nets [4] by CP-Nets with generalized additive models. UCP-nets are able to represent preferences quantitatively rather than representing simply preference ordering.

Similar to CP-nets, we firstly specify preferential dependency among attributes. Instead of specifying a total preference ordering over the values of each attribute according to their parents' values (conditions), we assign a real value (utility) for all values of each attribute. Utility function $u(X_1, X_2, ...X_n)$ is represented in Equation 1 where X_i is the ith attribute of outcome, U_i denotes parents of X_i and $f_i(X_i, U_i)$ represents a factor. Assume that our UCP-Net involves three factors f_1(Neighborhood), f_2(Price) and f_3(Parking Area, Neighborhood). The utility of an outcome is estimated as the sum of these factors.

$$u(X_1, X_2, ...X_n) = \sum_i f_i(X_i, U_i) \tag{1}$$

3 Proposed Heuristics

Most of the negotiation strategies [6, 9] work with quantitative preferences such as *utility functions*. However, it is desired for users to express their preferences qualitatively. Thus, we propose heuristics to use acyclic CP-Nets in negotiation while agents still negotiate with their strategies using quantitative information, *utility* (a real value between zero and one). To do this, we generate predicted utilities from a given CP-Net by applying our heuristics.

In our framework, a preference graph is induced from a given CP-Net while eliciting a user's preferences as a CP-Net. In this preference graph, each node denotes a possible outcome and each edge represents an improving flip. The direction of edges are ordered from less desired to more desired services. Therefore, the worst outcome will be placed at the top of preference graph (root node) whereas the leaf node holds the best outcome. For intermediate nodes, we only compare the nodes having a path from others. The nodes having no path to each other cannot be compared.

Figure 1 shows a preference graph induced from a CP-Net. The node (*Yes*, *Etiler*, *Low*) represents a low-priced apartment at *Etiler* having a parking area. There is an edge from (*No*, *Kartal*, *High*) to (*No*, *Kartal*, *Medium*). This means that an apartment with a medium price at *Kartal* not having a parking area is preferred over an apartment with a high price at *Kartal* not having a parking area.

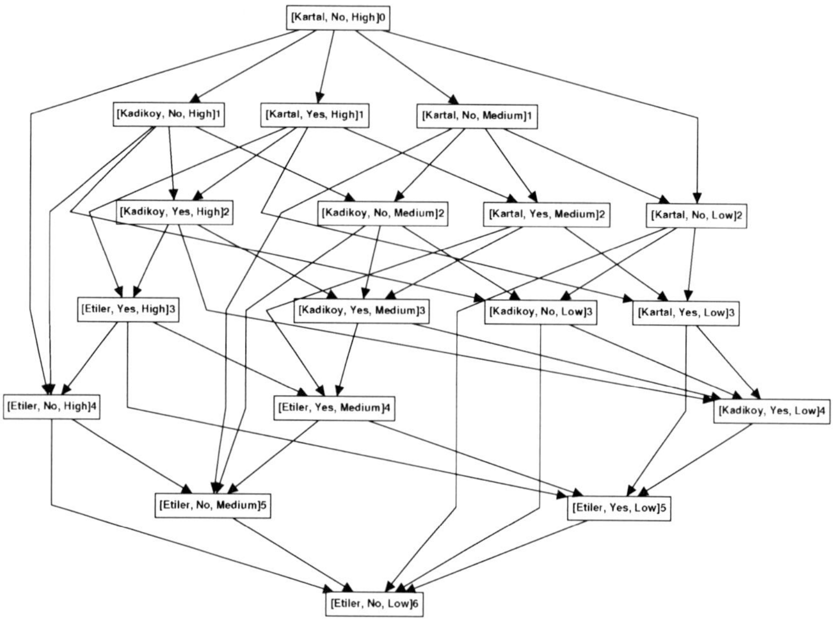

Fig. 1 Induced preference graph from a given CP-Net

An agent having a CP-Net applies one of the following heuristics and uses the estimated utilities produced by a chosen heuristic.

3.1 Depth Heuristic (DH)

We have previously proposed an approach based on capturing the depth of an outcome in preference graph [1] but in that study *depth* is used by the proposed negotiation strategy — it is not independent from the negotiation strategy. However, in this study we use the concept of *depth* to produce estimated utilities of outcomes regardless of negotiation strategy. That is, the agent using this heuristic is able to apply any negotiation strategy.

The depth of an outcome node in a preference graph indicates how far it is from the worst choice. It is intuitive to say that the better (more preferred) a service is, the further it is from the worst outcome. The depth of an outcome node is estimated as the length of the longest path from the root node keeping the worst choice.

According to this approach, the higher the depth of an outcome, the more likely it is to be preferred by the user. Further, if two outcomes are at the same depth, it is assumed that these outcomes are equally preferred by the user. We apply Equation 2 to estimate the utility values between zero and one. In short, the depth of a given outcome is divided by the depth of the preference graph (the highest depth) to obtain estimated utility of that outcome. For example, if we have a preference graph with a depth of 6 in Figure 1, an outcome whose depth is equal to 3 will have utility of $0.5 (= 3/6)$.

$$U(x) = \frac{Depth(x, PG)}{Depth(PG)} \tag{2}$$

3.2 Borda Scoring Heuristic

CP-Nets order outcomes partially and there are a plenty of linear orderings consistent with the partial ordering of outcomes induced from a CP-Net. One of these linear orderings may reflect the user's real preference orderings. Thus, this heuristic is based on finding all possible linear extensions of a given partial preference ordering and selecting one of the most suitable linear extensions.

One possible way of a linear ordering is to apply a voting procedure. To do this, we estimate all linear extensions of a given partial preference ordering induced from a preference graph and apply a voting procedure called "Borda Rule" [3] to obtain one of the most suitable linear orderings.

According to Borda Rule, we score outcomes according to their position in the ordering. Assume that we have m alternatives ordered as $< o_1, o_2 ... o_m >$ where o_{i+1} is preferred over o_i. When we score the outcomes, each outcome will get a point of its position minus one (o_i will get $i-1$). The sum of points namely *Borda count* represents the aggregation of existing alternative orderings. To illustrate this, consider we have three orderings such as $< x, y, z >$, $< z, x, y >$, $< x, z, y >$ where x, y and z are possible outcomes. Borda count of x would be equal to one $(= 0 + 1 + 0)$. In this approach, Borda count of each outcome over all possible linear extensions will reflect how much that outcome is preferred. Thus, we will estimate utilities based on the calculated Borda counts.

On the other hand, the number of all possible linear extensions of a given partial ordering may be so huge that this technique may become impractical because of high complexity. In order to reduce the complexity, we partition the preference graph and apply Borda Rule to all possible linear extensions of each subpartition.

How do we partition the preference graph? We know that the root node holds the worst outcome while the leaf node holds the best outcome. Thus, we need to find an ordering for the outcomes within the intermediate nodes. We partition this part in such a way that each subpartition can involve at most n, predefined number of outcomes. For this purpose, n can be taken as 10 or 15 according to the size of the preference graph. We choose 10 in this study. Figure 2 shows how we partition the preference graph in Figure 1.

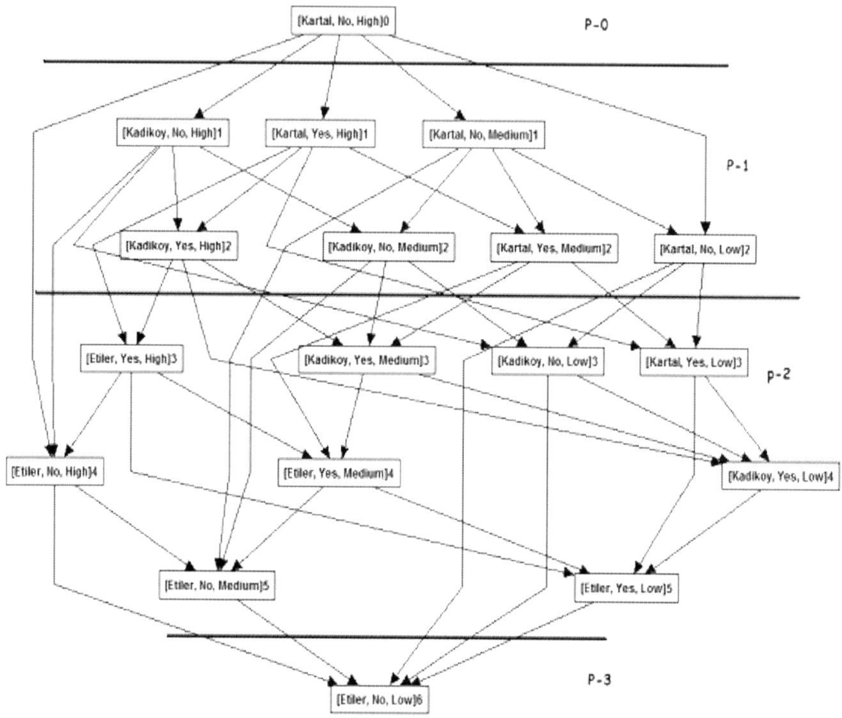

Fig. 2 Partitioning the preference graph in Figure 1

After applying Borda rule to each partition, we normalize Borda counts in a way that Borda count of each outcome will be between zero and one. To do this, we divide Borda count of each outcome in that partition by the maximum Borda count.

Another issue pertains to using these normalized Borda counts in order to estimate final utilities. We distribute the utilities by considering the number of outcomes at each partition. To achieve this, we apply the formula in Equation 3 where $U(x, p_i)$ denotes the utility of outcome x in the ith partition, $U_{max}(i-1)$ denotes the utility of outcome whose utility is maximum in the previous partition $(i-1)$, N denotes the number of possible outcomes, S_{p_i} denotes the number of outcomes in i^{th} partition and $BRCount(x, p_i)$ denotes normalized Borda count of the outcome x. $U_{max}(p_0)$, the utility of worst outcome (root node in the preference graph), is equal to $1/N$.

$$U(x, p_i) = U_{max}(p_{i-1}) + \frac{S_{p_i}}{N} * BRCount(x, p_i) \tag{3}$$

4 Experiments

To evaluate the proposed heuristics, we extend GENIUS [11], which is a platform for bilateral negotiation. Our extension enables an agent to elicit user's preferences as CP-Nets and to use utilities estimated by chosen heuristic while negotiating. The platform also stores the user's total ordering of outcomes as UCP-Nets and evaluates each negotiation outcome for that agent based on the given UCP-Net. The given UCP-Net is consistent with the given CP-net. In our experiments, the UCP-Net serves as ground truth. After an agent negotiates using its CP-Net, we evaluate its performance as if we knew the correct total ordering (UCP-Net).

We investigate three test cases to compare the performance of the heuristics. In each test case, two agents *Agent A* and *Agent B* negotiate with each other. We fix both agents' negotiation strategies so that *Agent A* negotiates with the same *Agent B* (having same preference profile and strategy). In the first case, *Agent A* has a CP-net and applies Depth Heuristic (DH) to derive the estimated utilities. During the negotiation, the agent will act on according to these estimated utilities. In the second case, *Agent A* has the same CP-Net with the first case but it applies Borda Scoring Heuristic (BSH) to estimate utilities which will be used in negotiation. In the last case, *Agent A* has its user's real total preference orderings in the form of UCP-Net (consistent with the CP-net and able to compare all outcomes). Thus, it uses the real utilities. Consequently, we are able to observe what the agent gets at the end of negotiation when it applies heuristics on partial preference information (CP-Net) versus when it has total preference information (UCP-Net).

In our experiments, each agent uses a concession based strategy in which the agent starts with the outcome having the highest utility and concedes over time. It also remembers the best counter offer that is made by the opponent agent. If the utility of the current counter offer is higher than or equal to the utility of agent's previous offer, then the agent will accept the offer. The agent will take the best counter offer of its opponent into account while generating its offer. If the utility of the current offer is lower than that of the best counter offer, the agent will take the opponent's best counter offer.

Since the opponent agent (*Agent B*)'s preference profile has a significant impact on negotiation outcome, we generate 50 different preference profiles for *Agent B*. That is, the same *Agent A* will negotiate with 50 different *Agent B*s. *Agent B*'s preferences are represented with a linear additive utility function in this experiment. Another factor having an influence on negotiation outcome in this setting is UCP-Net of the user. Different UCP-Nets mean different ordering of outcomes, so represent different users. Thus, we generate four different UCP-Nets for *Agent A* consistent with the given CP-net—four different users having the same CP-net. As a result, both agents will negotiate 200 times (4 different users of *Agent A* * 50 different *Agent B*).

Furthermore, we investigate the performance of the heuristics from a different point of view by taking the structure of CP-Nets into account. We generate three different CP-Nets. *CPNet*-1 involves one dependency such as preference of *parking area* depends on *neighborhood* whereas *CPNet*-2 involves two dependencies such as both preferences of *parking area* and *price* depend on *neighborhood*. There are not any dependencies between attributes in *CPNet*-3. For each CP-Net, we generate four different UCP-Nets consistent with them and perform the experiments mentioned above.

4.1 Sum of Utilities for Agent A

Our first evaluation criterion is the sum of negotiation outcomes' utilities with respect to *Agent A* over 50 different negotiations with *Agent B*. Table 1 shows these total utilities for three different CP-Nets and four different UCP-Nets consistent with each CP-Net. As expected *Agent A* using UCP-Net gets the highest score when it has a consistent UCP-Net with *CPNet*-1 and *CPNet*-3 since it negotiates with user's real preference orderings. Overall, the performance of the agent using BSH is quite close to that of the agent using UCP-Net (172 vs. 179 and 171 vs. 172). For the case of *CPNet*-2, the score of BSH is approximately the same as the score of UCP-net. Since *CPNet*-2 involves two dependencies (the user specifies her preferences in a more detailed way), the agent may get more information than the case of other CP-Nets (one dependency and no dependency). This leads to better results. The score of heuristics are the highest when they have *CPNet*-2.

Table 1 Sum of Outcome Utilities over 50 Negotiations for *Agent A*

AGENT A	DH	BSH	UCP-Net
CPNET-1 with UCPNet-1A	39.03	39.00	41.88
CPNET-1 with UCPNet-2A	38.27	40.73	43.66
CPNET-1 with UCPNet-3A	45.73	45.69	45.80
CPNET-1 with UCPNet-4A	46.88	46.94	47.29
Overall Sum (200 negotiations):	**169.91**	**172.36**	**178.63**
CPNET-2 with UCPNet-1B	39.93	41.66	41.70
CPNET-2 with UCPNet-2B	42.94	43.56	43.21
CPNET-2 with UCPNet-3B	46.15	46.76	46.75
CPNET-2 with UCPNet-4B	42.18	43.56	43.53
Overall Sum (200 negotiations):	**171.20**	**175.55**	**175.20**
CPNET-3 with UCPNet-1C	40.17	41.61	40.83
CPNET-3 with UCPNet-2C	41.58	42.50	45.64
CPNET-3 with UCPNet-3C	42.83	43.97	43.37
CPNET-3 with UCPNet-4C	42.36	43.36	42.64
Overall Sum (200 negotiations):	**166.94**	**171.44**	**172.48**

Moreover, *Agent A*'s score while applying Borda Scoring Heuristic (BSH) is higher than the case in which it uses Depth Heuristic (DH) for all CP-Nets (based on overall sum over 200 negotiations). According to this criterion, BSH may be preferred over DH.

4.2 Number of Times as Well as UCP-Net

Our second evaluation criterion is the number of times that the agent that applies a heuristic on a given CP-Net negotiates at least as well as the agent having a UCP-Net. If the utility of outcome for the agent using a heuristic is higher than or equal to the utility of outcome for the agent having UCP-Net, that agent receives one point. Since 50 different *Agent Bs* negotiate with the same *Agent A*, we evaluate this criterion over 50 negotiations.

According to Table 2, when *Agent A* uses *CPNet*-1 and applies DH, it negotiates at least as well as the agent having total preference ordering (UCP-Net) in 78 per cent of negotiations whereas BSH is successful at least as UCP-Net in 76 per cent of negotiations. Although the performance of BSH with respect to sum of utilities is better than that of DH, it negotiates as successfully as UCP-Net more than BSH for *CPNet*-1 (78 per cent versus 76 per cent). This stems from the fact that when BSH completes a negotiation better than DH, the difference between utilities of the outcomes is much higher than the case when DH negotiates better than BSH.

Table 2 Number of Times Heuristics Performs As Well As UCP-Nets

AGENT A	DH	BSH
CPNET-1 with UCPNet-1A	40	35
CPNET-1 with UCPNet-2A	26	35
CPNET-1 with UCPNet-3A	46	38
CPNET-1 with UCPNet-4A	44	44
Overall Sum (200 negotiations):	**156**	**152**
CPNET-2 with UCPNet-1B	43	49
CPNET-2 with UCPNet-2B	48	48
CPNET-2 with UCPNet-3B	44	50
CPNET-2 with UCPNet-4B	44	50
Overall Sum (200 negotiations):	**179**	**197**
CPNET-3 with UCPNet-1C	44	50
CPNET-3 with UCPNet-2C	27	31
CPNET-3 with UCPNet-3C	45	49
CPNET-3 with UCPNet-4C	47	47
Overall Sum (200 negotiations):	**163**	**177**

For *CPNet*-2 and *CPNet*-3, the agent using BSH negotiates successfully as the agent having UCP-Net more than the agent using DH. When agents have *CPNet*-2, it is seen that BSH beats DH. Note that in 89.5 per cent of negotiations DH negotiates at least as well as UCP-Net whereas 98.5 per cent of negotiations BSH performs at least as good as the UCP-Net.

5 Discussion

Our experimental results show that it would be better to apply Borda Scoring heuristic (BS) in small domains since its performance is higher than that of Depth heuristic (DH). However, we may prefer to use DH in large domains since its complexity is lower than BSH.

Li *et al.* study the problem of collective decision making with CP-Nets [10]. Their aim is to find a Pareto-optimal outcome when agents' preferences represented by CP-Nets. They firstly generate candidate outcomes to increase the computational efficiency instead of using the entire outcome space. Then each agent ranks these candidate outcomes according to their own CP-Nets. For ranking an outcome, they use *the longest path between the optimal outcome and that outcome* in the induced preference graph. Thus, the minimum rank is desired for the agents. They choose the final outcome for the agents by minimizing the maximum rank of the agents. In contrast, we use *the longest path between the worst outcome and that outcome* to estimate the utilities with our depth heuristic. Moreover, while they propose a procedure for collective decision making, we focus on estimating utility values of each outcome that will be used during the negotiation for an agent.

Rosi *et al.* extend CP-Nets to capture multiple agents' preferences and present *m*CP-Nets [12]. They propose several voting semantics to aggregate agents' qualitative preferences and to determine whether an outcome is preferred over another for those agents. They propose to rank an outcome in term of the length of the shortest sequence of worsening flips between that outcome and one of the optimal outcomes while we use the longest sequence of improving flips between the worst outcome and that outcome in our depth heuristic to get the estimated utilities.

Acknowledgements. This research is supported by Boğaziçi University Research Fund under grant BAP5694 and Turkish Research and Technology Council.

References

1. Aydoğan, R., Yolum, P.: Heuristics for cp-nets: Negotiating effectively with partial preferences. In: HUCOM Workshop in Group Decision and Negotiation, The Netherlands (2010)
2. Baarslag, T., Hindriks, K., Jonker, C., Kraus, S., Lin, R.: The First Automated Negotiating Agents Competition (ANAC 2010). In: Ito, T., Zhang, M., Robu, V., Fatima, S., Matsuo, T. (eds.) New Trends in Agent-Based Complex Automated Negotiations. SCI, vol. 383, pp. 113–135. Springer, Heidelberg (2012)
3. Black, D.: The Theory of Committees and Elections. Cambridge University Press (1958)
4. Boutilier, C., Bacchus, F., Brafman, R.I.: Ucp-networks: A directed graphical representation of conditional utilities. In: Proceedings of UAI 2001, USA, pp. 56–64 (2001)
5. Boutilier, C., Brafman, R.I., Domshlak, C., Hoos, H.H., Poole, D.: Cp-nets: A tool for representing and reasoning with conditional ceteris paribus preference statements. J. Artif. Intell. Res. (JAIR) 21, 135–191 (2004)
6. Faratin, P., Sierra, C., Jennings, N.R.: Negotiation decision functions for autonomous agents. Int. Journal of Robotics and Autonomous Systems 24(3-4), 159–182 (1998)

7. Fatima, S.S., Wooldridge, M., Jennings, N.R.: On Efficient Procedures for Multi-issue Negotiation. In: Fasli, M., Shehory, O. (eds.) TADA/AMEC 2006. LNCS (LNAI), vol. 4452, pp. 31–45. Springer, Heidelberg (2007)
8. Hindriks, K., Tykhonov, D.: Opponent modelling in automated multi-issue negotiation using bayesian learning. In: 7th International Joint Conference on Autonomous Agents and Multiagent Systems, AAMAS, Estorial, Portugal, pp. 331–338 (2008)
9. Jonker, C.M., Robu, V., Treur, J.: An agent architecture for multi-attribute negotiation using incomplete preference information. JAAMAS 15(2), 221–252 (2007)
10. Li, M., Vo, Q.B., Kowalczyk, R.: An efficient procedure for collective decision-making with cp-nets. In: Proceeding of ECAI 2010, Lisbon, Portugal, pp. 375–380 (2010)
11. Lin, R., Kraus, S., Baarslag, T., Tykhonov, D., Hindriks, K., Jonker, C.M.: Genius: An integrated environment for supporting the design of generic automated negotiators. Computational Intelligence (2012)
12. Rossi, F., Venable, K.B., Walsh, T.: mcp nets: Representing and reasoning with preferences of multiple agents. In: Proceeding of AAAI 2004, pp. 729–734 (2004)

An Implementation of Collective Collaboration Support System Based on Automated Multi-agent Negotiation

Mikoto Okumura, Katsuhide Fujita, and Takayuki Ito

Abstract. Recently, discussions among many people about global warming and global product development have been increasing. Efficient collaborative support based on multi-agent systems is necessary to collect the huge number of opinions and reach optimal agreements among many participants. We propose a collaborative park-design support system as an example of collective collaboration support systems based on multi-agent systems. In this system, agents elicit the utility information of users, collect many alternatives, and reach optimal agreements based on automated negotiation protocol. In particular, we focus on the steps for determining the attribute space and estimating the utility spaces of users in real world.

1 Introduction

Recently, discussions among many people about global warming and global product development are increasing. Efficient collaborative support based on multi-agent systems is necessary to collect huge number of opinions and reach optimal agreements among many participants. Many automated negotiation mechanisms are already existed. But the perfect utility functions of agents are assumed [1, 2, 3, 4, 5, 6, 7, 8]. In real world, it takes a lot of time to elicit the whole utility spaces of users.

We propose a collective collaboration support system based on the multi-issue automated negotiation mechanisms [1, 2, 3, 4, 5, 6, 7, 8]. In this system, the agents

Mikoto Okumura · Takayuki Ito
Nagoya Institute of Technology, Gokiso-cho, Showa-ku, Nagoya, Aichi, 466-8555 Japan
e-mail: okumura@itolab.nitech.ac.jp,
ito.takayuki@nitech.ac.jp

Katsuhide Fujita
The University of Tokyo, Kogaku bldg 9, room 136, 2-11-16, Yayoi, Bunkyo-ku, Tokyo, 113-0032, Japan
e-mail: fujita@ipr-ctr.t.u-tokyo.ac.jp

T. Ito et al. (Eds.): Complex Automated Negotiations, SCI 435, pp. 125–141.
springerlink.com

elicit the utility information of users, collect many alternatives, and reach optimal agreements based on the automated negotiation protocol. Especially, we focus on the steps of deciding the attribute space and estimating the utility spaces of users in real world.

We adopt a collaborative park-design support system as an example of a collective collaboration support system. Many users, like citizens and designers, should join the work to design parks. Many opinions and preferences of participants should be respected. Additionally, the designs of parks have some interdependent issue, for example, there are some dependence between the amount of playground equipments and the cost. In such a case, the automated negotiation protocol with issue-interdependency is effective [5, 6, 7, 8]. But applying an automated negotiation protocol with issue-interdependency requires utility functions of users because most of the papers assumed the perfect utility functions. In real world, it is impossible to elicit all the utility information of users.

Our system estimates the interdependent multi-attributes utility functions of users based on users' evaluations of the designs generated by our system. The utility function is composed of some simple fundamental functions. One fundamental function is defined by one user's evaluation of design. The fundamental functions has a character that the utility grows low as the point is far from the sampling point corresponding to the design. The bumpy utility space is generated by combining the simple mound of the fundamental functions.

We report our works as follows. First, we describe the outline of the collaborative park-design support systems. Next, we propose a new method of estimating the utility functions of users in real life. Third, we demonstrate some results of our method and conduct an experiment to evaluate the effectiveness of our system. Finally, summarizing our research.

2 Collaborative Park Design Support System

Fig. 1 shows the outline of a collaborative system based on multi-agent system. The details of the system are followings.

`[Step1]` **Collecting the opinions and preferences**
The system decides sampling points and generates the alternative of the park design at the sampling point. After that, this system elicit the uses' preferences based on users' evaluations.

`[Step2]` **Estimating utility functions**
The system predicts the whole utility spaces based on the sampling points collected in the `[Step1]`. This estimation will be shown in the section4.

`[Step3]` **Multi-agent automated negotiations**
The agent which is behalf of the users finds the optimal agreements by the automated negotiation protocol[1, 2, 3, 4, 5, 6, 7, 8].

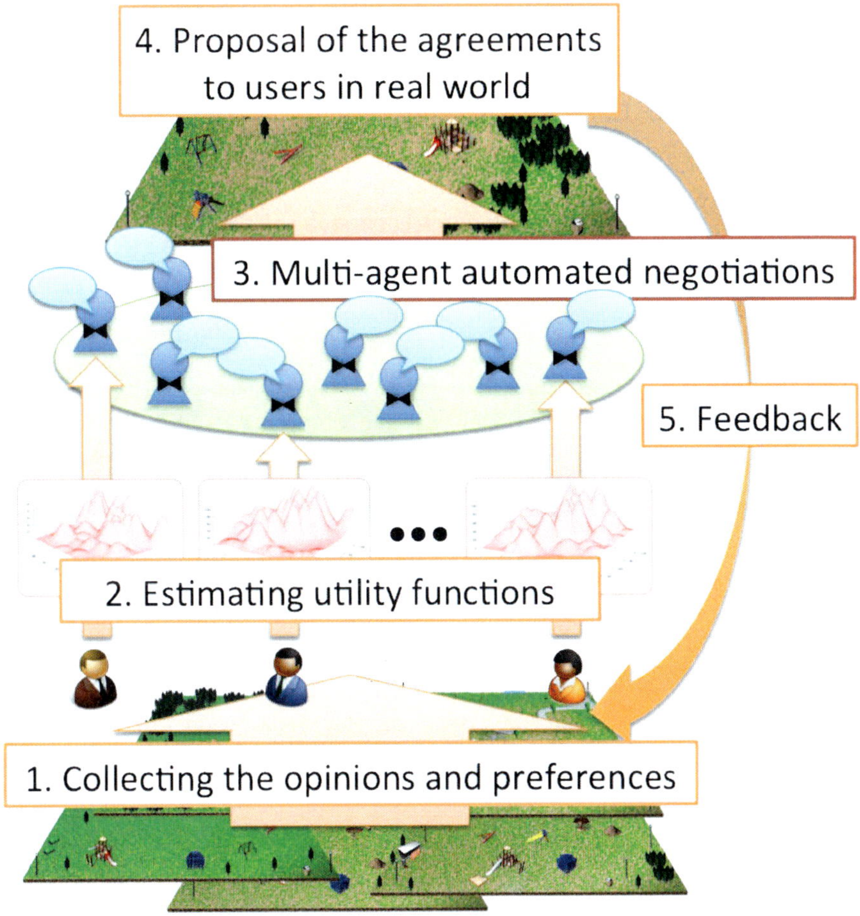

Fig. 1 Collaborative public space design processes

`[Step4]` **Proposal of the agreements to users in real world**
The system generates a design from the agreement (a point on the attribute space) in `[Step3]`.

`[Step5]` **Feedback**
The system sends the design result(final alternative) in this round, and the users give a feedback. If the most of users agree to the final alternative, it is an optimal agreement. If the most of users don't agree to the final alternative, these steps are repeated(`[Step1]`~`[Step5]`). But, our system doesn't have this step to be simple because our study and implementations of system based on multi-agent system are in its early stage.

3 Constraints of a Feasible Attribute Space

We specify constraints of an attribute space to implement a system in real world. In this section, M is a set of models which are targets of negotiation. In our system, a model is a design of park. And A is an attribute space which is a set of attribute vectors. U is a utility function.

The first constraint is existence of the function f in the expression(1). The function f maps a model to an attribute vector. This constraint is needed to translate a negotiation target to a computational form which can be understood by an agent.

$$f : M \to A \tag{1}$$

The second constraint is existence of the function g in the expression(2). The function g maps an attribute vector to a model. This function g is needed to translate a negotiation result produced by agents to a model which can be understood by a human.

$$g : A \to M \tag{2}$$

The third constraint is the function f and g satisfy the expression(3). In a process of negotiation, a model is mapped to an attribute vector by the function f, agents negotiate on an attribute space using mapped models and a produced result which is an attribute vector is mapped to a model by the function g. This fact requires an utility of a model is similar to an utility of a translated model in a negotiation process. In other words, this constraint(3) means the consistency between a model and an attribute vector through a process of negotiation.

$$U(x) \simeq U(g \cdot f(x)),\ x \in M \tag{3}$$

In our system, the constraint(4) which is more strict than the constraint(3) is employed to simplify an implementation. Employing a parametric model and using parameters as attributes are guarantee to existences of the function f and g, satisfying the constraint(3). Additionally, a parametric model allows to automatically generate models for acquiring many evaluations of models.

$$x = g \cdot f(x),\ x \in M \tag{4}$$

4 A Method of Eliciting the Utility Spaces

The method of eliciting utility functions corresponds to the `[Step1]`,`[Step2]` in the section2. This system generates the park designs automatically, receives the users' evaluations, and estimates the utility spaces.

The utility function is composed of some simple fundamental functions. The fundamental functions has a character that the utility grows low as the point is far from the sampling point. The bumpy utility space is generated by combining the simple mound of the fundamental functions.

4.1 Fundamental Function

Definition 0.1. Fundamental Function

$\mathbb{R}^+$ is a set of positive real numbers more than 0, $\mathbb{R}^*_+$ is a set of all positive real numbers. When i is an index, s_i shows a sampling point, d_i is the distribution of f_i and v_i is the evaluation value of $s_i(v_i, d_i \in \mathbb{R}^*_+)$. The fundamental function f_i is defined as a following expression.

$$f_i(\mathbf{x}) = v_i \cdot \exp(-\frac{(\mathbf{x} - \mathbf{s_i})^2}{d_i}) \tag{5}$$

- The fundamental function is always more than 0 and a multi-dimensional space.
$$f_i : \mathbb{R}^{+^n} \to \mathbb{R}^+$$
- The maximum of the fundamental function is equal to the evaluation value of the user.
$$\max f_i(\mathbf{x}) = v_i$$
- The maximum point of fundamental function means the sampling point.
$$\arg\max_{\mathbf{x}} f_i(\mathbf{x}) = \mathbf{s_i}$$
- The value of the fundamental function is smaller as it grows far from the sampling point.
$$||\mathbf{x_1} - \mathbf{s_i}|| > ||\mathbf{x_2} - \mathbf{s_i}|| \to f_i(\mathbf{x_1}) < f_i(\mathbf{x_2})$$

4.2 Combining the Fundamental Functions

Definition 0.2. Utility Function

When there are N sampling points; $(\mathbf{s_1}, .., \mathbf{s_N})$, the utility function U is defined as follows:

$$U(\mathbf{x}) = \max_{i=1,..,N} f_i(\mathbf{x}) \tag{6}$$

But the definition 2 has two problems as Fig. 2, Fig. 3 shows. In the left part of Fig. 2, the sampling s_j don't work well because the function f_j is totally smaller than the function f_i. For instance, our system should employ f_j at the square area because the sample point of f_j (s_j) is closer to the square area than that of f_i (s_i). In the left part of Fig. 3, the sampling s_j don't work well because most of the function f_j is smaller than the function f_i. For instance, our system should employ the function

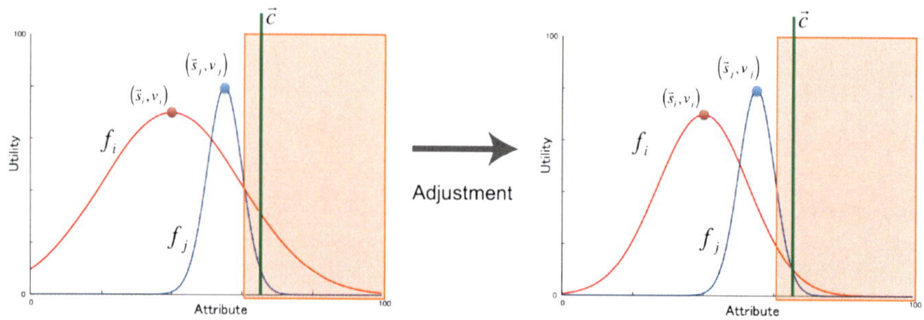

Fig. 2 The Fundamental Function is Under the Other Fundamental Function (Case1)

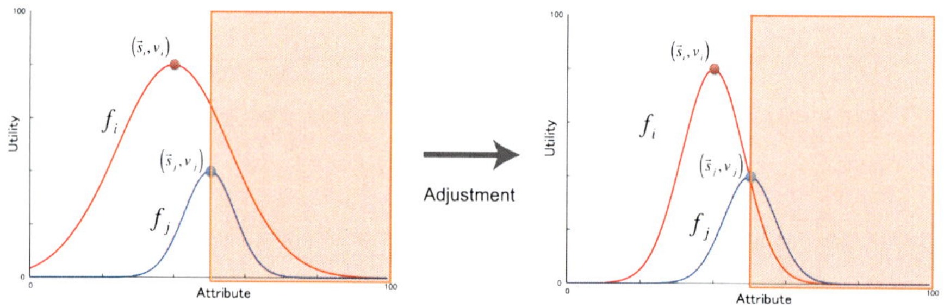

Fig. 3 Most of the Fundamental Function are Under the Other Fundamental Function (Case2)

f_j at the square area in the Fig. 2 because the point in the square area is closer to the sampling point of function f_j (s_j) than that of function f_i (s_i). Following two techniques resolve these two problems by modifying d_i which is the distribution of the fundamental function f_i.

Method 1. *A fundamental function is under other fundamental function (Case 1)*

This method adjusts the f_i as Fig. 2 showing by modifying d_i. For instance, we assume that two different sampling points $\mathbf{s_i}, \mathbf{s_j}((i \neq j, \max f_i(\mathbf{x}) \geq \max f_j(\mathbf{x})))$ exist. If $f_i(\mathbf{s_j}) > f_j(\mathbf{s_j})$, then this method modifies d_i to satisfy $f_i(\mathbf{s_j}) = f_j(\mathbf{s_j})$ using the expression(7).

$$d_i = \frac{(\mathbf{s_j} - \mathbf{s_i})^2}{\ln \frac{v_i}{v_j}} \tag{7}$$

Method 2. *Most of a fundamental function are under other fundamental function (Case2)*

This method adjusts the f_i as Fig. 3 showing by modifying d_i. For instance, we assume that two different sampling points $\mathbf{s_i}, \mathbf{s_j}(i \neq j, d_i > d_j)$ exist. If $f_i(\mathbf{c}) > f_j(\mathbf{c})$, then this method modifies d_i to satisfy $f_i(\mathbf{c}) = f_j(\mathbf{c})$ by the expression(8).

$$d_i = \frac{(\mathbf{c} - \mathbf{s_i})^2}{\frac{k^2}{2} - \ln \frac{v_j}{v_i}} \tag{8}$$

$\mathbf{c} = \mathbf{s_j} + k\sqrt{\frac{d_j}{2}}\mathbf{u}$, $\mathbf{u} = \frac{1}{||\mathbf{s_j} - \mathbf{s_j}||}(\mathbf{s_j} - \mathbf{s_j})$. $\mathbf{u}$ is the unit vector whose direction is from $\mathbf{s_i}$ to $\mathbf{s_j}$. $\mathbf{c}$ is a control point of adjusting. $\mathbf{c}$ depends on the parameter $k(\in \mathbb{R}^*_+)$. As k grows, this method is performed at the point which distance from s_j is large. For example, $\mathbf{c}$ goes right as k grows in Fig. 3.

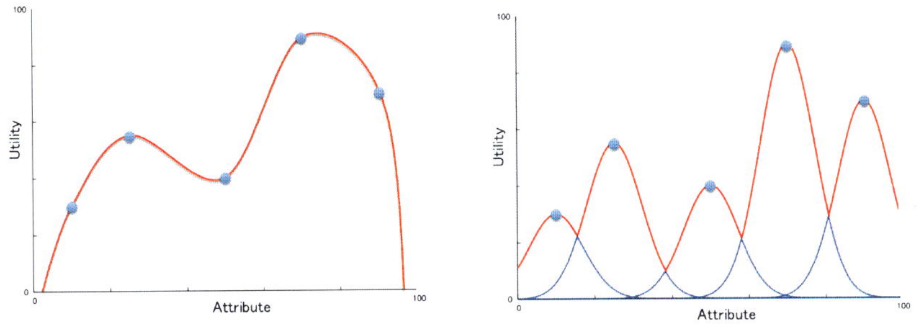

(A) Connecting Sampling Points Smoothly

(B) Employing Maximum Points of Fundamental Functions Directory (Our method)

Fig. 4 A Estimated method of Connecting Sampling Points Smoothly and Employing Maximum Points of Fundamental Functions Directory

The simple way of estimating utility functions is to connect all sampling points smoothly as Fig. 4 showing. But the utility is usually estimated as higher value than real one when the distance between some sampling points is large. Our method improves this issue by using a maximum of fundamental functions as Fig. 4 showing.

Our method has a tendency to make agreement at the area of containing more information because the utility with enough information is large. By contract, it is difficult for agents to make agreement when the number of samples is not enough. But our method modifies this problem because the sampling points are decided based on the users' suggestions.

4.3 Estimating User's Utility Space

Our system estimates the user's utility space as follows(`[Step1]`‘`[Step4]`).

`[Step1]` **Creating a Fundamental Design**
The manager of the negotiation sets up a negotiation. The manager creates a fundamental design by the user-interface as Fig. 5 showing. The manager decides the arrangements of trees, playground equipments, facilities and so on. The manager can check the park designs generated automatically by our system, and change some parameters for reflecting his ideas.

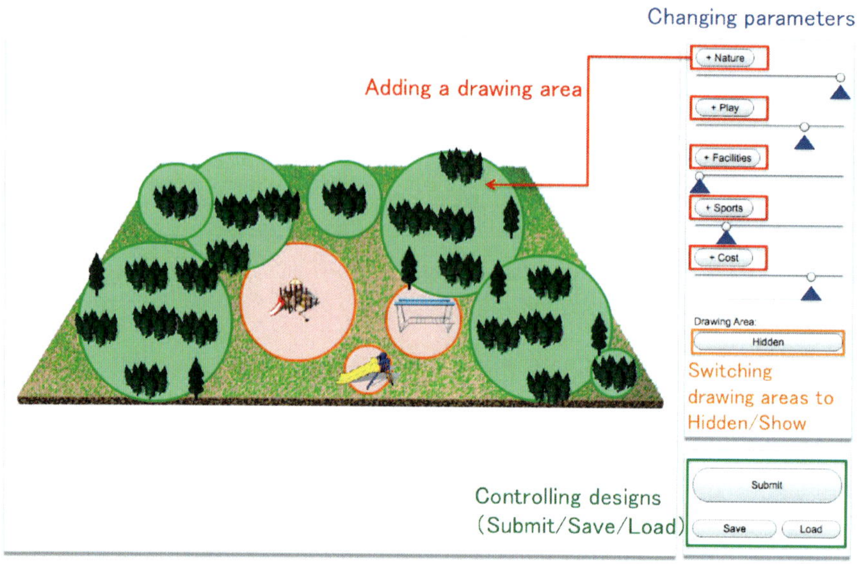

Fig. 5 User Interface of Creating a Fundamental Park Design

`[Step2]` **Deciding Sampling Points**
Our system decides some sampling points in the attribute space. In our system, the sampling points are selected randomly.

`[Step3]` **Evaluation by the users**
Our system generates the park designs at the sampling points. There are some appraising methods for evaluating the sampling points (e.g. voting, rating). We employ the rating method. Users rate each park design and submit the results of rating by the user-interface as Fig. 6 showing.

`[Step4]` **Estimating Utility Functions**
First, the system generates the fundamental functions. Next, our system combines all of the fundamental functions by Method1 and Method2). Specifically, these methods adjust $d_i, d_j (0 \leq j < i)$. d_i is initialized by $D_0 (d_i := D_0$, D_0 is the initial value of the distribution of fundamental function).

`[Step2]`∼`[Step4]` is repeated during the period of negotiation decided by the manager in the [Step1].

5 An Example of Estimating a Utility Function

In this section, we demonstrate some results of our systems. The purpose of the demonstration is to evaluate our method and to show the characters of our method.

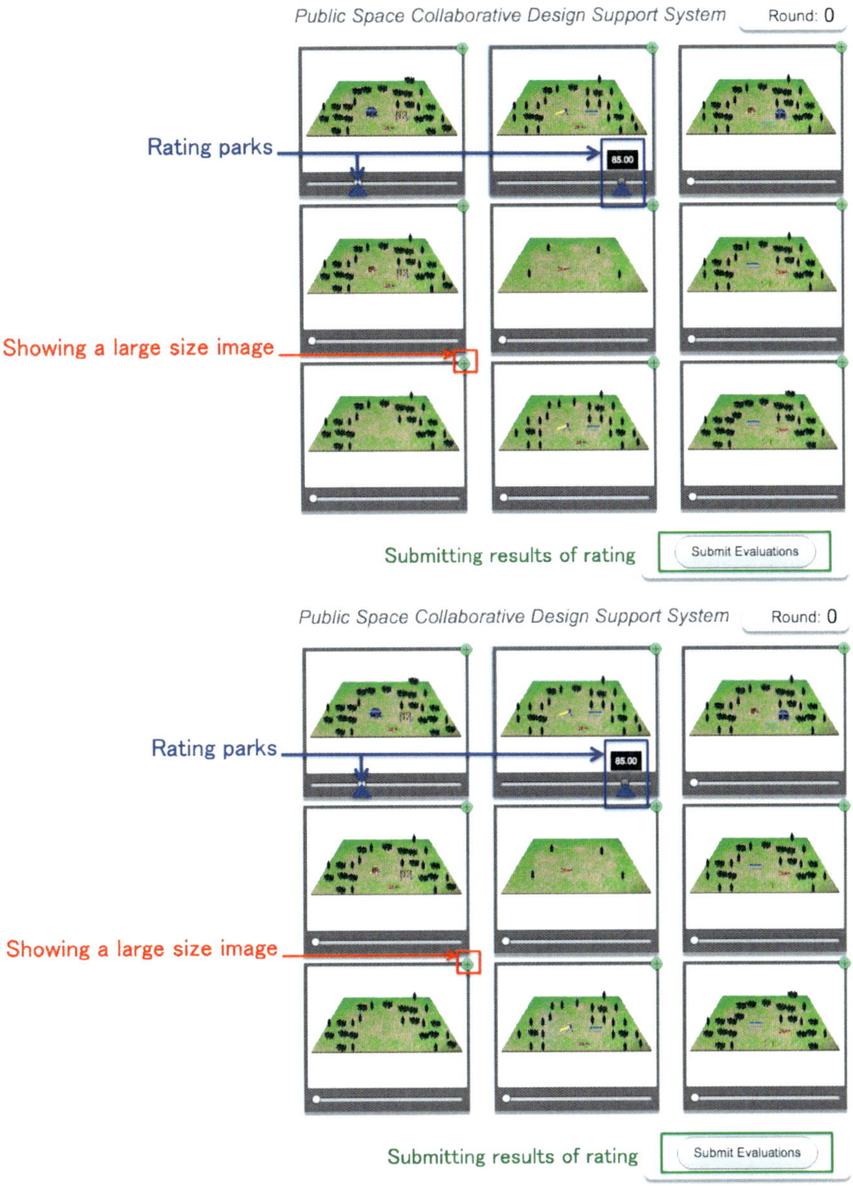

Fig. 6 User Interface of Evaluating Park Designs

In this demonstration, we assume that the user has a following idea: "The parks which have many trees and some playground equipments are good. The parks which have too many or few playground equipments are not so good." Fig. 7 shows an example of the estimated utility function. In this demonstration, the number of

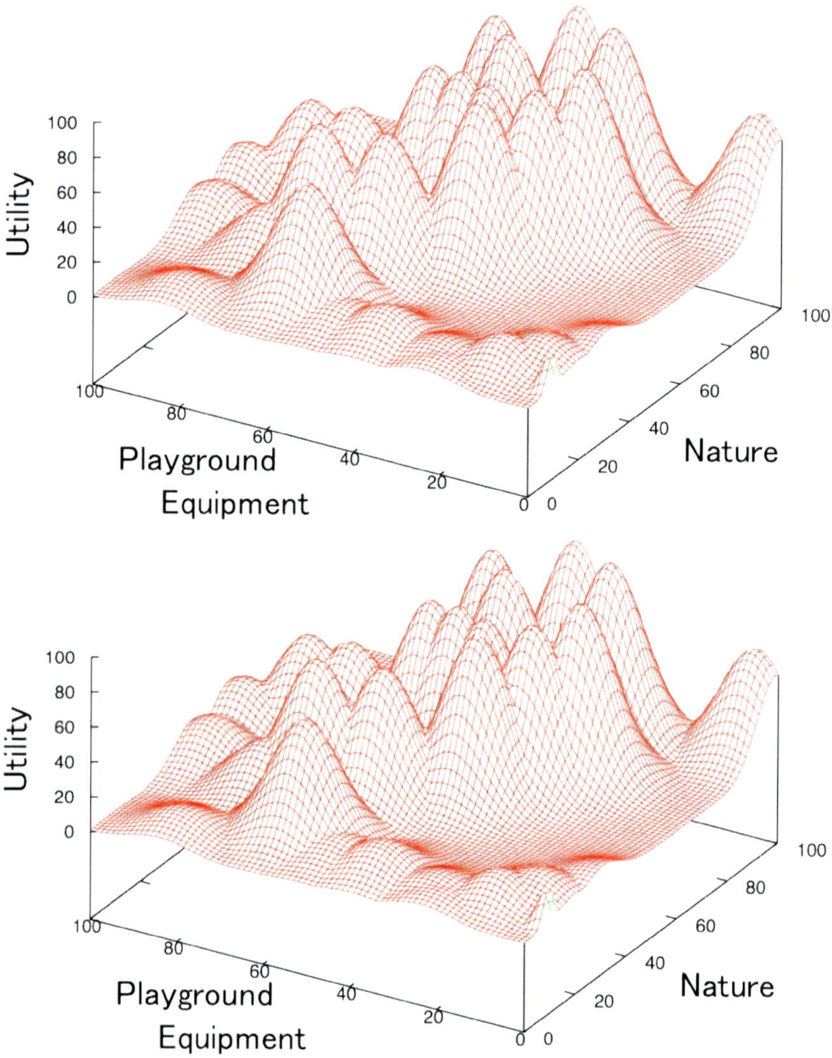

Fig. 7 Estimated Utility Function

sampling is 30 and the number of attributes is 2. The reason of small number of attributes is that we can't show graphically when the number of attributes is more than 3. As you know, our method can be applied when the number of attributes is more than 3.

The axis "Nature" in Fig. 7 shows how rich the nature of the park is. The large value of "Nature" means that the park has rich nature. The axis "Playground Equipment" shows how many the playground equipments are in the park. The large value of "Playground Equipment" means that the park has many playground equipments.

In Fig. 7, the utility is the highest when "Playground Equipment" is 50 and "Nature" is the high value. Therefore, the estimated utility function represents the accurate preferences of the user. But the utility is too high when "Nature" is 60 and "Playground Equipment" is 50. This is because that too many samplings are happened at the point. The efficient sampling for estimation of utility spaces is one of the future work.

6 Experiments

6.1 Setting of Experiments

We conducted an experiment to evaluate the effectiveness of our system and confidence of our preference elicitation method. In the experiment, we ran 2 negotiations. In each negotiation, a number of participants is 11. "Nature" and "Playground Equipment" are used as attributes. Each attribute is a real number which is bigger than or equal to 0 and less than or equal to 100. The period of first negotiation is 10 minutes and second negotiation is 5 minutes. Because participants understood our system, the period of second negotiation is reduced. To find the optimal agreement, we used simulated annealing (SA) and the best result of 5 SAs is adopted as the optimal agreement because SA is easy to implement and finding optimal agreement is not our main work. After the negotiations, we send out questionnaires to get users' comments.

6.2 Experimental Results

Fig. 8 shows the results of negotiations. Table 1 shows some information about utility values for the results. U1 and U2 are user's rates (real utility values) for the results. Est1 and Est2 are user's estimated utility values for the results. Err1 and Err2 are margins of error between real utility value (U1, U2) and estimated utility value (Est1, Est2). U1, Est1 and Err1 are values for the first negotiation. U2, Est2 and Err2 are values for the second negotiation.

An average of U1 is 79.09 and U2 is 79.36. Therefore, we find many people agree to the results. An average of Err1 is 15.78 and Err2 is 16.60. These values are not so good but our method of preference elicitation can elicit tendencies of user's preference.

Table 1 shows that most of the users' utility function are accurately elicited like User5 but some users' preference elicitation are not accurately elicited like User3. Fig. 9 is a elicited utility function of User5. This case is preference of a user is

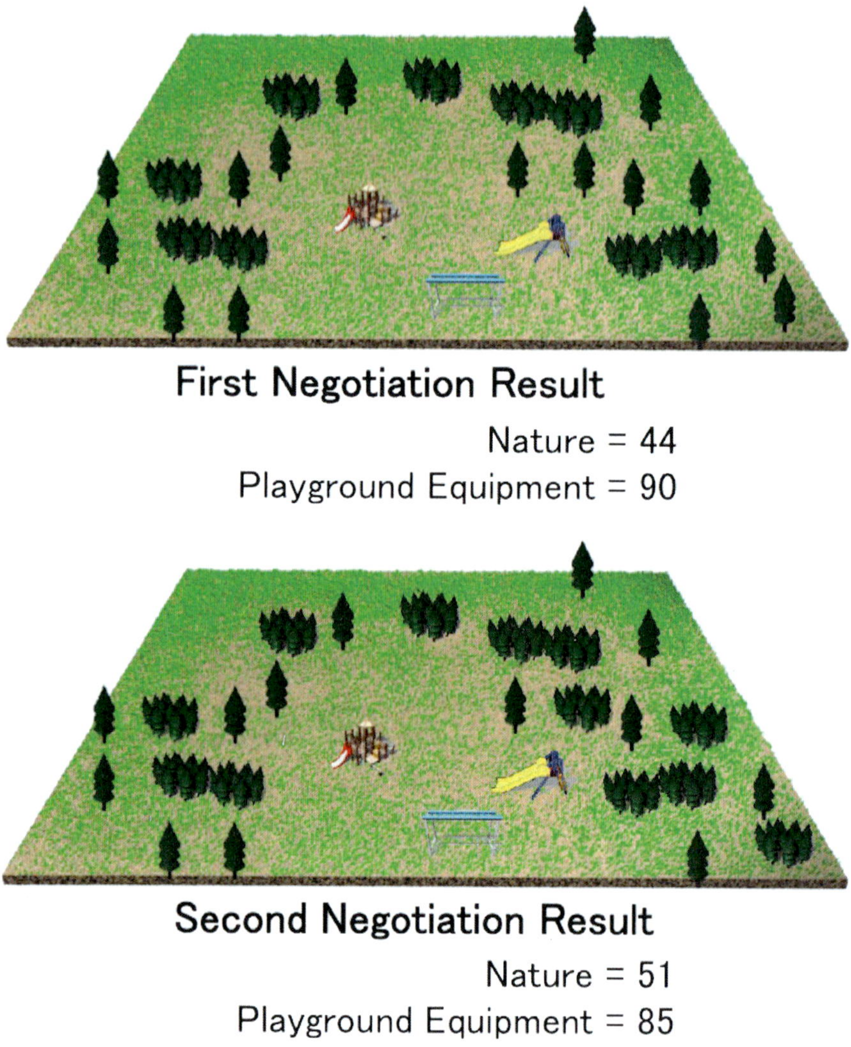

Fig. 8 Negotiation Results

accurately elicited. Fig. 10 is a elicited utility function of User3. This case is preference of a user is not accurately elicited. A shape which has many sharp mounds like Fig. 10 occurs when some close points on a attributes space have very different utility values, in other words, some similar park designs are got very different rates. A reason of this problem is no considering changes of preferences. In fact, User3 commented "My criteria of rating are inconsistent in a process of evaluations." on a questionnaire. In real world, human preferences change as time passes. Additionally, almost all methods of preference elicitation[9, 10, 11] and utility theory has this

Table 1 Utility Values for The Optimal Agreement

	U1	U2	Est1	Est2	Err1	Err2
User1	80	85	94	61	14	24
User2	80	80	-	64	-	16
User3	60	50	19	74	41	24
User4	80	85	-	-	-	-
User5	85	75	80	73	5	2
User6	90	97	62	82	28	15
User7	70	80	57	51	13	29
User8	90	86	67	58	23	28
User9	80	80	90	87	10	7
User10	90	90	86	75	4	15
User11	65	65	69	71	4	6
Average	79.09	79.36	69.30	69.60	15.78	16.60

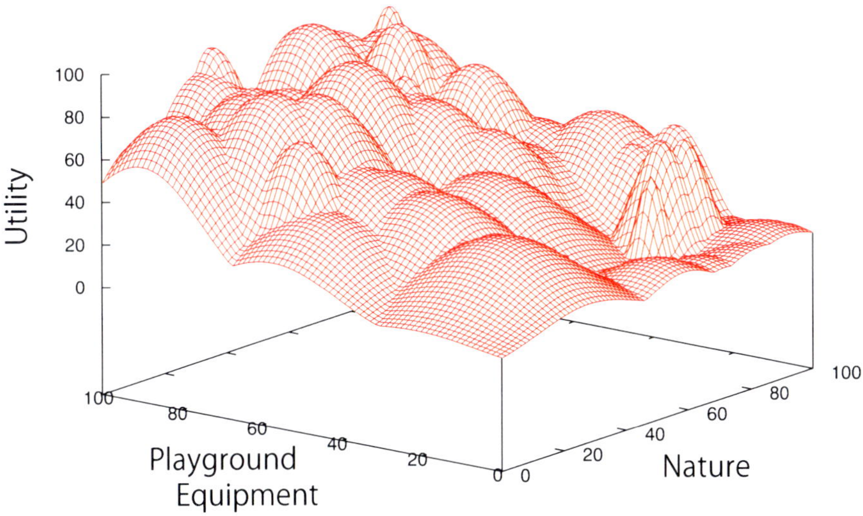

Fig. 9 A Good Case of Elicitation (Elicited Utility Function of User5)

problem, too. It seems establishing "time" as a attribute resolves this problem but sampling points on a utility space from a user can not be finished. Because a number of samples getting at once is limited and samples acquired on different steps have different time as an attribute. As a result, a number of samples can not be enough to describe a utility space.

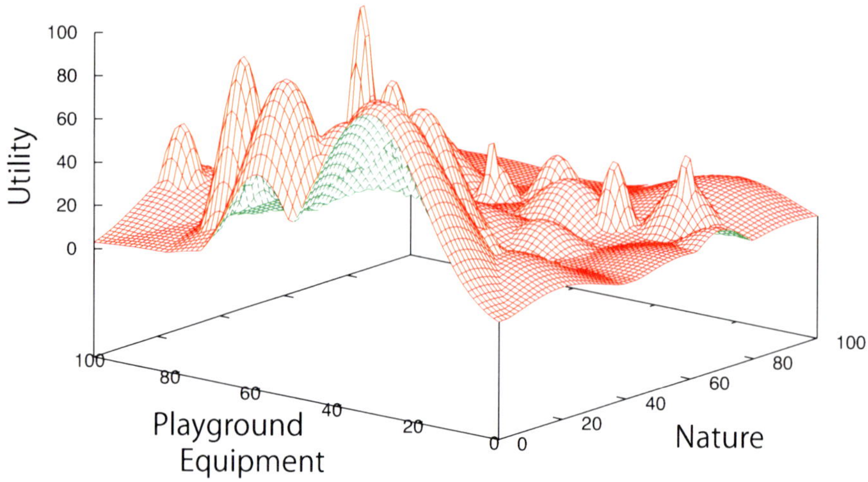

Fig. 10 A Bad Case of Elicitation (Elicited Utility Function of User3)

7 Related Works

Most previous works on multi-issue negotiation have addressed only linear utilities [1, 2, 3, 4]. Recently some researchers have been focusing on more complex and non-linear utilities. For example, Ito, Fujita and Mizutani et. al[5, 6, 7, 8] proposes the automated negotiation protocol with issue-interdependency. But most of the paper assumed the perfect utility functions of agents. In real world, it is impossible to elicit the all utility information of agents. We propose the method of estimating the utility spaces with issue-dependences.

Luo et al.[9] proposes a method of eliciting and quantifying the trade-off between issues by the user-interactions. But the system don't work well when the utility function is complex with the dependences between more than 3 issues. On the other hand, our system can work well when the utility functions are more complex.

In [12, 13], the system supports a yard-design based on interactive GA. This system can generate the efficient yard-designs based on the preference of users. But this system isn't assumed the multi-party negotiations. Our system supports to the multi-party collaborative designs and consensuses among many users.

8 Conclusion

We implemented a collaborative park-design support system based on multi-agent systems. In particular, we focused on the steps for determining the attribute space and estimating the utility spaces of users in real world. First, we formulated the constraint (3) that must be satisfied by an attribute space so that it can be used to implement real world application. As a result, we proposed a use of parametric designs for a feasible system as one of compromised points which satisfies the constraint (4) which is more strict than the constraint (3). Second, we proposed a method of eliciting user's utility space. At last, our experimental results shows our system succeeded to build a consensus.

We have three future works. First, current utility theories do not consider changes of human preferences. By analyzing experimental results, we found some changes affecting negotiation results. Resolving this problem requires a new theory. Second, current preference elicitation methods and negotiation protocols do not consider differences of density among attributes. In real world, some attributes are more important or sensitive for changes of attribute values than other attributes. Resolving this problem requires both of a new preference elicitation method and a new negotiation protocol. Resolving this problem, negotiation time will be reduced and correctiveness of negotiations result will grow At last, improvements of system are required to apply practical problem. For example, implementation of feedback is required to build better agreements and a better method of selecting sampling points than random sampling is required for more correct estimation of utility spaces.

Appendix

The Production of the Expression (3)

The expression (7) modifies d_i to satisfy $f_i(\mathbf{s_j}) = f_j(\mathbf{s_j})$.

$$f_i(\mathbf{s_j}) = f_j(\mathbf{s_j})$$

$$v_i \cdot \exp(-\frac{(\mathbf{s_j} - \mathbf{s_i})^2}{d_i}) = v_j$$

$$d_i = \frac{(\mathbf{s_j} - \mathbf{s_i})^2}{\ln \frac{v_i}{v_j}}$$

d_i is produced.
To f_i becomes the gaussian, d_i must be positive.

$$d_i = \frac{(\mathbf{s_j} - \mathbf{s_i})^2}{\ln \frac{v_i}{v_j}} > 0$$

$$v_i > v_j$$

Because of $f_i(s_j) > f_j(s_j)$ which is the condition of applying the method 1,

$$v_i = f_i(s_i) > f_i(s_j) > f_j(s_j) = v_j$$

$d_i > 0$ is evidenced.

The Production of the Expression (4)

The expression (8) modifies d_i to satisfy $f_i(\mathbf{c}) = f_j(\mathbf{c})$.

$$f_i(\mathbf{c}) = f_j(\mathbf{c})$$
$$v_i \cdot \exp(-\frac{(\mathbf{c}-\mathbf{s_i})^2}{d_i}) = v_j \cdot \exp(-\frac{(\mathbf{c}-\mathbf{s_j})^2}{d_j})$$

Because of $\mathbf{c} = \mathbf{s_j} + k\sqrt{\frac{d_j}{2}}\mathbf{u}$,

$$\exp(-\frac{(\mathbf{c}-\mathbf{s_i})^2}{d_i}) = \frac{v_j}{v_i} \cdot \exp(-\frac{k^2}{2}\mathbf{u}^2)$$

$\mathbf{u}^2 = 1$ because $\mathbf{u}$ is a unit vector.

$$\exp(-\frac{(\mathbf{c}-\mathbf{s_i})^2}{d_i}) = \frac{v_j}{v_i} \cdot \exp(-\frac{k^2}{2})$$
$$d_i = \frac{(\mathbf{c}-\mathbf{s_i})^2}{\frac{k^2}{2} - \ln\frac{v_j}{v_i}}$$

d_i is produced.

References

1. Faratin, P., Sierra, C., Jennings, N.R.: Using similarity criteria to make issue trade-offs in automated negotiations. Artificial Intelligence 142, 205–237 (2002)
2. Shaheen Fatima, S., Wooldridge, M., Jennings, N.R.: Multi-issue negotiation with deadlines. J. Artif. Intell. Res. (JAIR) 27, 381–417 (2006)
3. Shaheen Fatima, S., Wooldridge, M., Jennings, N.R.: Approximate and online multi-issue negotiation. In: AAMAS, p. 156 (2007)
4. Shaheen Fatima, S., Wooldridge, M., Jennings, N.R.: An analysis of feasible solutions for multi-issue negotiation involving nonlinear utility functions. In: AAMAS (2), pp. 1041–1048 (2009)
5. Ito, T., Hattori, H., Klein, M.: Multi-issue Negotiation Protocol for Agents: Exploring Nonlinear Utility Spaces. In: Proc. of IJCAI 2007, pp. 1347–1352 (2007)
6. Fujita, K., Ito, T., Klein, M.: Approximately fair and secure protocols for multiple interdependent issues negotiation. In: AAMAS (2), pp. 1287–1288 (2009)
7. Fujita, K., Ito, T., Klein, M.: Secure and efficient protocols for multiple interdependent issues negotiation. Journal of Intelligent and Fuzzy Systems 21(3), 175–185 (2010)
8. Mizutani, N., Fujita, K., Ito, T.: Issue Clustering and Distributed Genetic Algorithms for Multi-issue Negotiations. In: ACIS-ICIS 2010, pp. 593–598 (2010)
9. Luo, X., Jennings, N.R., Shadbolt, N.: Acquiring user strategies and preferences for negotiating agents: A default then adjust method. International Journal of Human Computer Studies 64(4), 304–321 (2006)

10. Koichi, K., Yutaka, I.: Estimation of the Environmental Value of Recycled Wood Wastes: A Conjoint Analysis Study. Journal of Forest Research 5(1), 1–6 (2000)
11. Sandholm, T., Boutilier, C.: Preference Elicitation in Combinatorial Auctions. In: Proceedings of the Third ACM Conference on Electronic Commerce (2001)
12. Nanba, M., Kobe, M., Cai, D.: Japanese Garden Design using Interactive GA. IPSJ SIG Notes of Information Processing Society of Japan (IPSJ) (96), 55–62 (2001)
13. Nanba, M., Kobe, M., Cai, D.: Designing Japanese Garden using Interactive GA. IPSJ SIG Notes of Information Processing Society of Japan (IPSJ) (86), 53–58 (2003)

A Qualitative Ascending Protocol for Multi-issue One-to-Many Negotiations

Liviu Dan Şerban, Cristina Maria Ştefanache,
Gheorghe Cosmin Silaghi, and Cristian Marius Litan

Abstract. Many practical distributed systems environments require novel automatic mechanisms including multi-attribute reverse auctions for efficient partner selection and contracts negotiation. Recent results [6] show that the property of transferable utilities is not of vital importance, as qualitative versions of the standard auctions (e.g. qualitative Vickrey auctions (QVA), qualitative English auctions (QEA)) are proved to exhibit nice efficiency properties as well. Such auctions require that the preferences of the auctioneer are publicly known. However, practical protocols of multi-bilateral closed negotiations between a buyer and multiple sellers are experimentally shown [9] to approximate the Pareto-efficient best-seller QVA outcome, without requesting that any of the parties explicitly reveals their preferences. The only condition is to enable bidders to learn preferences. In this paper we introduce two novel negotiation protocols that approximates a qualitative ascending English auction (QEA) and overcomes some restrictions imposed by the non-transferable utilities environment. Our auction-like protocols are designed for fully automatic environments with learning agents playing the bidders' roles.

1 Introduction

Classical auctions (especially English and Vickrey) are widely studied in the literature and existing theoretical results can guide a mechanism designer to select the proper protocol for trading services in a distributed environment. In the context of independent private-value models, we highlight that the English and Vickrey auctions exhibit several important properties [12]: (i) the two auctions are strategically

Liviu Dan Şerban · Cristina Maria Ştefanache · Gheorghe Cosmin Silaghi ·
Cristian Marius Litan
Babeş-Bolyai University,
Str. Theodor Mihali 58-60, 400591, Cluj-Napoca, Romania
e-mail: {liviu.serban, cristina.stefanache}@econ.ubbcluj.ro
gheorghe.silaghi, cristian.litan@econ.ubbcluj.ro

T. Ito et al. (Eds.): Complex Automated Negotiations, SCI 435, pp. 143–159.
springerlink.com

equivalent, (ii) the outcome of both is Pareto-optimal and (iii) the winner is the bidder who values the object most highly. In the English auction the dominant strategy of a player is to bid actively until the price announced by the auctioneer reaches the value of the object to him. In the Vickrey auction, if a bidder knows the value of the object to himself, the dominant strategy is to submit a sealed bid equal to that value. These observations are valid in auction models with *transferable utilities*, i.e. a good designated as a currency is established in the community, bids are expressed in the form of price quotations and the auctioneer simply prefers a higher price to a lower one. From these fundamental assumptions, auctions evolved to environments where a center holds some amount of money, being interested to buy the best good or service he can get for that amount.

In this paper we consider such a reversed auction setup, with one buyer against several sellers, while bids are expressed as bundles of characteristics of the good / service under discussion. Such setups are of interest in (e.g.) service-based computing environments, where service level agreements (SLAs) are designated as contracts between providers and consumers, comprising quality of service parameters, price paid and penalty for violation of QoS parameters [1]. SLAs can be either individually negotiated between a provider and a consumer [18] or auctions can be employed to select the provider that supplies the best SLA. Nowadays, automatic SLA partner selection and negotiation is a hot topic in distributed computing.

To address this reversed auction setup, several novel practical and theoretical results exist. Harrenstein et al. [6] defines a *qualitative Vickrey auction* (QVA) protocol that generalizes the classical second-price Vickrey auction for environments without payments. They show that the dominant strategy of each bidder is to make that offer, which among the ones that are acceptable to him, is most preferred by the center. If all bidders adhere to this strategy, the outcome is weakly Pareto-optimal. They also suggest that in an English-like auction the straightforward strategy for each bidder is to offer the highest alternative in his preference order, such that the new bid is preferred by the center to the last submitted bid. In both cases the results are derived under the main assumption that the center publicly announces his preference profile at the beginning of the auction. Considering this issue and referring to the QVA setup of [6], Hindriks et al. [9] provides a practical protocol that approximates the QVA outcome.

Within the research framework just described, this paper introduces a protocol to practically implement a qualitative ascending English auction (QEA), overcoming the restrictions imposed by the non-transferable utilities environment. In our enterprise, we bypass the limitation that the buyer is unable to explicitly elicit his value function [3] or he might not know the complete domain of possible outcomes of the sellers [9]. For fully automatic environments, we provide an auction-like protocol that is expected to enable learning agents playing the bidders' roles, and we investigate whether the QEA outcome can be approximated in practical experiments. Given that the master does not reveal his profile and only signals some information to the competitors and the protocol does not have a termination deadline, the auction can extend to a huge number of rounds, letting each competitors to learn the profiles of the others. Thus, the setup transforms in a public-like one, and we expect the master obtain a better outcome than than in a private value auction, as suggested by [16]. Imposing a deadline to the negotiation, the setup moves closer to a private-value negotiation and we expect the outcome to better approximate the theoretical QEA one.

The structure of the paper is the following. Section 2 defines the multi-issues auction model with non-transferable utilities, including the definition of the theoretical qualitative English auction. Section 3 presents two practical protocols implementing the qualitative English auction with learning agents on the bidders' side, while the auctioneer's preferences (utility function) is not publicly known. Section 4 presents and discusses the experimental results. Section 5 briefs other approaches regarding concurrent negotiation. Section 6 concludes the paper.

2 Definitions

We tackle with a virtual environment where sellers and buyers negotiate over a good or service. Each service has m issues of interest $x = (x_1, x_2, \dots, x_m)$ with $x \in X = X_1 \times X_2 \times \cdots \times X_m$. For a service-based environment, these issues define functional or non-functional quality of service values, the penalty or the price of a service level agreement [1]. Buyers and sellers associate an utility value $u_i(x) \in [0,1]$ for each outcome x, and each player i has a reservation value v_i below which he does not accept any outcome. The utility functions u_i can be written as linear combinations of the individual utility functions $u_{i,k}$: $u_i(x) = \sum_{k=1}^{m} w_k u_{i,k}(x_k)$, with $\sum_{k=1}^{m} w_k = 1$, where $u_{i,k}(x_k)$ represents the utility that the agent i (buyer or seller) obtains by receiving the value x_k for the issue k and $0 \le w_k \le 1$ represent the weights measuring the importance of a given issue k for the agent. Agents prefer a higher utility to a lower one. An outcome x is *weakly Pareto-efficient* if there is no other outcome under which all players are strictly better off.

Further in this paper, without loss of generality, we shall assume a reverse-auction setup with one buyer and many sellers. The goal of the buyer is to design a mechanism that provides an efficient outcome and which is best possible for him. Formally, the buyer (with the utility function u_0) is interested in selecting the seller i^* such that:

$$i^* = \arg\max_{i \in \{1,\dots,n\}} \max\{u_0(x) | x \in X, u_i(x) \ge v_i\} \tag{1}$$

As in [6, 9], the above-defined environment is called a *qualitative* one, since no general accepted currency is defined and the bids and outcomes are represented as vectors x.

The classical English auction as described by [12, 6] implies that the center announces the price level (or the acceptable bid in a qualitative setup) and bidders adhere or not to the center's announcements. Variants of the classical English auction exist and among them, the one in which the bidders themselves call the bids and the auction ends when no one is willing to raise the bid. In such an English auction, in the case of independent private-values, the dominant strategy of a bidder is to always bid a small amount over the current highest bid and to stop when his private value is reached [16]. The equivalence between this sort of English auction and the second-price auction holds in a transferable-utility environment with private-value agent models [12]. In non-private English auctions with at least three bidders, the auctioneer is better off than in the Vickrey case, because other bidders' willing to raise the price causes any bidder to increase his own valuation of the item [16].

Listing 3. The qualitative English auction

1 The buyer announces his preferences.
While only one bidder remains or no bidder is willing to raise the bid utility of the buyer during a time frame
2. Bidders submit offers with a higher utility for the buyer than the best offer announced by the other bidders.
3. At any time, a bidder can withdraw (without a re-entering possibility)
End while
4. The last remaining bidder wins with his last bid.

Given these theoretical assertions, we define the qualitative English auction protocol as in Alg. 3. In QEA, the buyer accepts only bids in increasing order (of the global ordering induced by his preferences) until no bidder is interested any more to submit. Therefore, a straightforward seller strategy is to offer the highest alternative in his preference order, that it is higher (in the global ordering of the buyer) than the last submitted bid [6]. On the other hand, in the QVA protocol described in [6], the sellers' dominant strategy is to bid an offer that is just acceptable to himself and ranks highest in the buyer's preferences. Furthermore, QVA selects the best seller and let him to adjust his offer up to the utility of the second best seller.

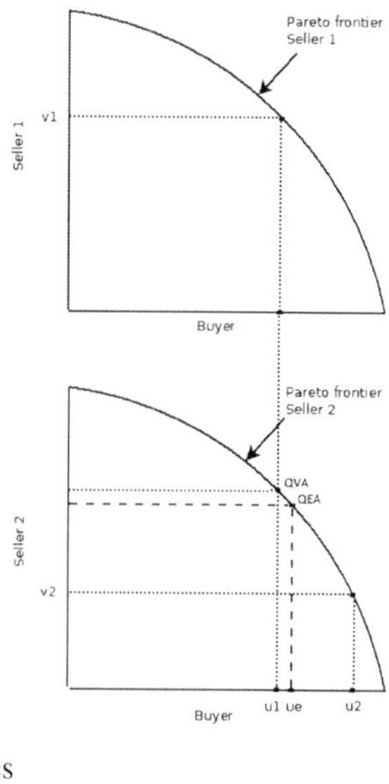

Fig. 1 QEA and QVA outcomes

Fig. 1 presents a comparison between the QEA and the QVA outcomes. Let a buyer asking for an outcome x and let us consider two sellers with reservation values $v_1 > v_2$. In both auctions, seller 2 wins, because it can supply better outcomes for the buyer. In the first round of the QVA, seller 2 announces a bid which is the closest (or identical if possible) to u_2 for the buyer (and to v_2 for himself). This bid is preferred by the auctioneer to the bid offered by seller 1. In the second round, seller 2 is allowed to adjust his bid up to u_1 (the utility for the buyer induced by seller 1's offer). Thus, the game will end in the outcome denoted by QVA.

On the other side, in QEA both sellers will announce bids, giving increasing utility to the buyer. When the outcome reaches the point (u_1, v_1), the first seller will exit, as he is not able to further improve his bid. Thus, only seller 2 remains in the game with the offer assessed with utility u_e for the buyer, i.e. representing the next offer ranked highest in seller 2's preferences, but with increasing utility for the buyer.

We note that the QEA outcome is the next outcome on the Pareto frontier for the above winning seller such that the utility of the buyer is higher than the one induced by the QVA outcome. If the domain is continuous, the utility u_e of the QEA outcome is only ε-better ($0 < u_e - u_1 < \varepsilon$ for a very small $\varepsilon > 0$) than the one of the QVA outcome and practically, both outcomes coincide. However, if the domain is discrete (and sparse) and the Pareto frontier is defined by discrete points, a significant difference between the utility of the QEA outcome and the one of the QVA may occur. Therefore, in the discrete case, a center willing to improve his outcome, might select the QEA.

3 Approximating the QEA

In this section we present two practical protocols that tries to practically implement the QEA. In the theoretical QEA (Alg. 3), to submit bids with increasing utility for the buyer, sellers need to know the buyer's preferences or the buyer's utility function. If, because of various reasons, the buyer cannot reveal his preferences (nor his utility function, if preferences have such representation) or the buyer might not be able to define all possible outcomes of the game, another signaling procedure should be used, such that the sellers to be notified to deliver increasing value bids to the buyer.

3.1 The 1st Protocol (Withdraw Condition)

This protocol, described in Alg. 4, consists of several rounds of concurrent negotiation sessions between the buyer and each of the sellers. In this case, we do not consider a time limit by which the interactions must end. With several sellers, the ending condition is met when all but one of the sellers withdraw from the negotiation because they do not accomplish the condition of the while statement. The seller who remained up to the end wins with his last offer and signs the agreement with

Listing 4. The 1st concurrent negotiation protocol (withdraw condition)

1. Each seller i sets $U_{0,i}^{exp} = 0$

2. While sellers i have offers $x^{(i)}$ such that $u_{0,i}^{exp}(x^{(i)}) > U_{0,i}^{exp}$ and $u_i(x^{(i)}) \geq v_i$, for those sellers i perform:

2.1 Each seller submits an offer $x^{(i)}$ with expected buyer utility higher than $U_{0,i}^{exp}$

2.2 The buyer selects the offer $x^{(i^*)}$ with the highest utility $u_0(x^{(i^*)})$

2.3 The buyer announces the bid $x^{(i^*)}$ to each seller

2.4 Each seller updates its $U_{0,i}^{exp}$ to the expected utility value for the bid $x^{(i^*)}$

end while

3. The last remaining bidder wins with the her last bid

the buyer. We believe that through this protocol with multiple rounds, the buyer obtains an optimal outcome x^* for him. In this scenario, it is sufficient that the buyer to reveal only partial information regarding her preferences, i.e. the highest bid at every round.

With incomplete information, the sellers shall adopt a learning based strategy to predict the consumer's preferences, which is an essential condition for them to be able to produce increasing value bids. We assume that each seller *learns* the buyer's profile from his bids, exactly like in an one-to-one bilateral negotiation. Initially, all buyer's profiles from sellers' point of view are blank, thus, each seller initializes to zero the threshold $U_{0,i}^{exp}$ for the expected utility for the buyer. In each round, each seller i who has not reached his reservation value v_i, submits a bid $x^{(i)}$ such that this bid is better off for the buyer: $u_{0,i}^{exp}(x^{(i)}) > U_{0,i}^{exp}$. More precisely, for a rational seller, this bid will be exactly the next one in decreasing order, following his preference ordering, which increases the expected utility for the buyer. If, for a given seller i, there are no more bids which satisfy the condition above, that particular seller should withdraw from the negotiation.

In each round, the buyer collects all the expressed bids and announces the best bid $x^{(i^*)}$ for him. We notice that the buyer reveals little information, he just signals the best submitted bid to all sellers. Each seller, who is capable of learning the opponents' profiles, will infer both the buyer's and the best competitor's profile, exactly like in an English auction, where one sees all the other bids. Further, the round winning seller will know that he won the round, because his bid was repeated by the buyer.

To keep the similarity with the QEA, we translated the exiting condition of line 3 in Alg. 3 in the `while` condition of Alg. 4. This condition states that a seller will remain in the protocol while he has offers, acceptable to him, that are greater than the threshold of the round. If, within a round, a seller does not submit an offer, this means that the seller exited the auction.

One possible problem with this protocol is that each bidder should submit an offer in a reasonable amount of time. In this first version of the protocol, we assume that the sellers are not trying to exploit the time restrictions, thus, if a seller has an available offer, it will present this offer, if not, he will immediately withdraw.

A similar behavior is obtained if we impose a short time limit for every round. If a seller does not respond with an offer during the time limit, he is considered withdrawn from the negotiation.

Note that the protocol proposed here has many nice practical properties. First, it does not require the buyer to make public his profile. Furthermore, even if the buyer does not a priori know the range of the possible offered outcomes, this does not represent a problem for the protocol, as the buyer is not required (at any time) to produce an outcome. All what buyer is requested is that he is able to rank the possible incoming offers. Second, the protocol is ascending - simulating an English auction; at every round sellers being presented the highest bid up to the moment. Third, the protocol can simulate a non-private value setup, the sellers being able to learn the competitors' profiles by observing the best bid in each round. Thus, the protocol might yield higher utility for the buyer.

The existence of learning agents is essential for the success of the protocol presented in Alg. 4. Learning agents usually perform two steps when producing an offer:

- process all information from the environment to infer a model of the environment, auctioneer and the competitors;
- select and propose the next offer.

We consider that such a learning agent can be easily adapted to the protocol presented in Alg. 4, because from the seller agent's point of view, he plays a game like in the one-to-one negotiation. For our experiments presented in section 4 we extended the Bayesian learning agents [9] to be compatible with the rules of this concurrent protocol.

In the next subsection, we present the second protocol developed for implementing the QEA, where we included another ending condition, namely a deadline for the negotiation. We will see that in the presence of such a condition, the agents should adapt their bidding strategy to reach a favorable agreement in a reasonable time.

3.2 The 2nd Protocol (Deadline Condition)

The main motivation to development this second concurrent protocol are the drawbacks encountered in the case of the first protocol, namely the time needed for reaching a final agreement between the buyer and one of the sellers. In Alg. 4, a seller withdraw only when he exhausted and proposed all his available bids, thought that are better than the previous round best one. Thus, for huge domains with thousands of bid possibilities, this would lead to a very high number of rounds, and therefore, a long time to reach an agreement.

To overcome this drawback and keep the practicality of the protocol, in Alg. 5 we included a deadline T for the whole negotiation session. The rules of the game are pretty much the same as those described in Alg. 4.

Listing 5. The 2nd concurrent negotiation protocol (deadline condition)

1. Each seller i sets $U_{0,i}^{exp} = 0$
2. While sellers i have offers $x^{(i)}$ such that $u_{0,i}^{exp}(x^{(i)}) > U_{0,i}^{exp}$ and $u_i(x^{(i)}) \geq v_i$ or a deadline T is not reached, for those sellers i perform:
2.1 Each seller submits an offer $x^{(i)}$ with expected buyer utility higher than $U_{0,i}^{exp}$
2.2 The buyer selects the offer $x^{(i^*)}$ with the highest utility $u_0(x^{(i^*)})$
2.3 The buyer announces the bid $x^{(i^*)}$ to each seller
2.4 Each seller updates its $U_{0,i}^{exp}$ to the expected utility value for the bid $x^{(i^*)}$
end while
If a deadline T is not reached:
3. The last remaining bidder wins with the her last bid
Else
4. All sellers i gain 0 from this negotiation session

Including the time limit T, requires all seller agents to adapt their bidding strategy to this. Notice that if no agreement is reached at the end of the time interval T, all agents get 0 utility.

ANAC 2010 competition[1] proved that, imposing a time limit to bilateral negotiating agents will make those agents to exploit this constraint as an asset [2]. Therefore, majority of agents devised for that competition avoided to accept agreements at the beginning of the negotiation time and majority of the agreements were accepted when the time limit was about to vanish. The Bayesian learning agent we used to test the 1st protocol with only a withdraw condition was adapted for a time constrained bilateral negotiation by dividing the negotiation time into intervals and let the agent to perform concessions only at the end of the negotiation [17]. With this idea in mind, we modified the Bayesian learning agent for the one-to-many negotiation protocol of Alg. 5 as follows:

- To fit the `while` condition of the protocol, the seller agent will select the next offer (at round $r+1$), such that $u_{0,i}^{exp}(x^{(i)}) > U_{0,i}^{exp}$;
- The agent will consider a concession factor β, which depends on the time t elapsed from the start of the negotiation. When selecting the bid, the agent will decrease its round utility target such us: $u_i(x_{r+1}^{(i)}) = u_i(x_r^{(i)}) - (\frac{t}{T})^{\frac{1}{\beta}}$.

As suggested in the general framework of [17], we developed a strategy that considers a differentiated concession factor β on various time intervals.

In the next section, we analyze the efficiency of the two concurrent negotiation protocols presented above, by comparing their outcomes with those selected by the theoretical QEA.

[1] The Automated Negotiating Agents Competition 2010 held on May 12, 2010, with the finals being run during the AAMAS 2010 conference http://mmi.tudelft.nl/negotiation/index.php/ANAC_2010

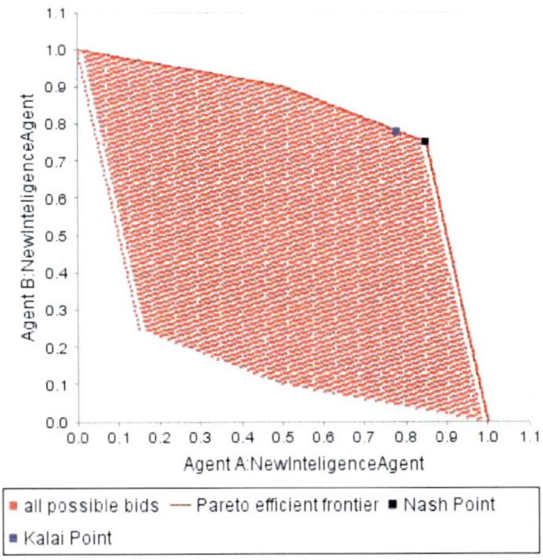

Fig. 2 The SON domain

4 Experiments and Results

In this subsection we present the experimental setup used for evaluating the proposed negotiation protocols. To assure comparability of results, we used the same negotiation domain as in [9]. For the scope of this paper, we considered only two sellers. In future, we will extend the results to many sellers, as the simulation software permits this.

To simulate the negotiation, we used a modified version of the Genius simulator [7] by extending the interaction layer to allow *one-to-many* negotiations.

As indicated in [9], the negotiation domain should be generic, to further ensure the generality of the results. Service-oriented negotiation domain described in [5] fits these requirements. This domain refers to negotiating a service with four issues of interest: delivery time, quality, duration and penalty, each issue having 30 possible values. In this domain, there are 810000 possible agreements. Figure 2 depicts this domain with all the possible bids. We notice the Pareto frontier and the Kalai-Smorodinsky point with utility values $(0.73, 0.73)$. Reaching agreements on the Pareto frontier is a desired solution, because no player can further improve without decreasing the utility of the opponents. Further, we would like the outcome to lay in the vicinity of the Kalai-Smorodinsky point, because it induces the fairness[2] property in the game.

Adapted for the one-to-many negotiation, we used the 12 preference profiles per role that were considered in [9]. The profiles combine various reservation values

[2] A Kalai-Smorodinsky outcome represents the Pareto efficient result for which the utilities of both players are proportional with their maximum possible gains. Thus, given that a total utility $u_1 + u_2$ is produced out of a game, the Kalai-Smorodinsky solution enforces an equalitarian (fair) distribution of this utility between the two players.

Table 1 Possible preference profiles for the buyer

Preference profile	w_1	w_2	w_3	w_4	Evaluation function	RV
Buyer1	0.3	0.15	0.2	0.35	ascending	0.3
Buyer2	0.5	0.3	0.05	0.15	ascending	0.4
Buyer3	0.45	0.2	0.1	0.25	descending	0.5
Buyer4	0.55	0.05	0.3	0.1	descending	0.65
Buyer5	0.15	0.45	0.35	0.05	triangular	0.75
Buyer6	0.65	0.1	0.05	0.2	triangular	0.8
Buyer7	0.15	0.25	0.1	0.5	ascending	0.2
Buyer8	0.1	0.2	0.55	0.15	ascending	0.35
Buyer9	0.2	0.55	0.15	0.1	descending	0.45
Buyer10	0.15	0.2	0.05	0.6	descending	0.55
Buyer11	0.05	0.6	0.25	0.1	triangular	0.6
Buyer12	0.25	0.15	0.05	0.55	triangular	0.7

Table 2 Possible preference profiles for the seller

Preference profile	w_1	w_2	w_3	w_4	Evaluation function	RV
Seller1	0.4	0.2	0.15	0.25	ascending	0.3
Seller2	0.2	0.35	0.3	0.15	ascending	0.4
Seller3	0.5	0.3	0.15	0.05	descending	0.5
Seller4	0.3	0.15	0.35	0.2	descending	0.6
Seller5	0.15	0.35	0.4	0.1	triangular	0.7
Seller6	0.6	0.05	0.15	0.2	triangular	0.8
Seller7	0.4	0.3	0.2	0.1	ascending	0.55
Seller8	0.2	0.55	0.1	0.15	ascending	0.2
Seller9	0.6	0.25	0.1	0.05	descending	0.15
Seller10	0.05	0.1	0.25	0.6	descending	0.25
Seller11	0.1	0.6	0.05	0.25	triangular	0.35
Seller12	0.5	0.1	0.05	0.35	triangular	0.45

and weights for each issue of the service. In Table 1 and in Table 2 we illustrate the preference profiles created for the buyer and the seller, respectively.

Playing one buyer against two sellers results a total of 792 possible games. We conducted all our experiments on a sample of size 50 out of the full population of 792 possible auction games.

4.1 The 1st Negotiation Protocol against QEA

To work with the Genius simulator, we adapted the Bayesian learning agent [8] for playing the protocol described in Alg. 4 as seller. This adapted Bayesian agent will never accept a bid proposed by the buyer and will withdraw when no further possible bids are available. The condition to select the next bid is further restricted by considering for selection only bids that have estimated utility for the buyer higher than the buyer's last announced bid. When winning the round, the Bayesian agent

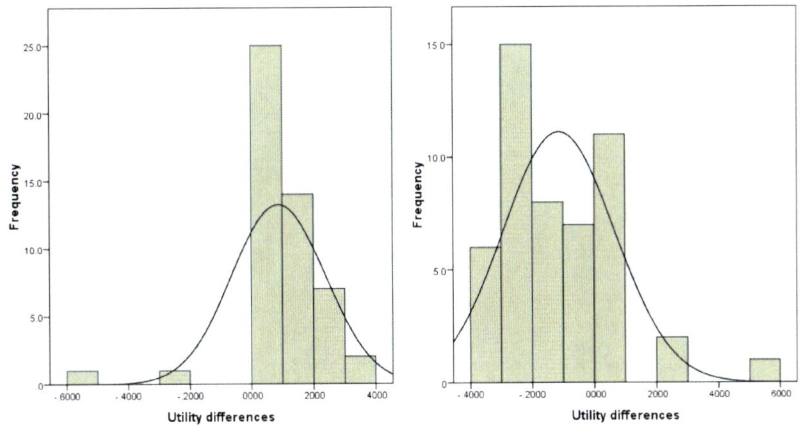

Fig. 3 Utility difference between the first concurrent negotiation protocol and the QEA on the buyer side (left) and on the seller side (right)

Table 3 Synthetic statistical results for utility differences between the first concurrent negotiation protocol and the theoretical QEA

Agent	Mean	Std. Dev.	95% conf. interval Lower	Upper
buyer	0.087150	0.149194	0.044749	0.129551
seller	-0.113400	0.177679	-0.163896	-0.062904

will repeat the winning bid, up to the moment that another competitor submits a better bid.

The simulations reveal that in 100% of the cases, the winner selected by the negotiation mechanism coincides with that selected in the *Qualitative English Auction*(QEA).

In Fig. 3 we present the distributions of the utility differences between the first protocol (described in Alg. 4) and QEA for the buyer and the winning seller[3]. Table 3 presents the 95% confidence internal the for mean values. In the case of the buyer, we can see that there is a slightly left asymmetry. Also, the average utility difference is $\overline{U}^{b}_{diff} = 0.087150$ which means that in the case of the first negotiation protocol we obtain a slightly higher utility than in the QEA. On the other hand, for the winning seller, the distribution of the utility differences is characterized by a slightly right asymmetry and the average utility difference is $\overline{U}^{s}_{diff} = -0.113400$ which means that we obtained a lower utility than in the QEA. Both confidence intervals for the utility difference of the buyer and the utility difference of the winning seller do not contain the value zero, clear indicating that the buyer is better of than in the theoretical QEA and the winning seller worse off.

[3] In the figures presented above, the solid line represents the standard normal distribution with mean zero and variance 1.

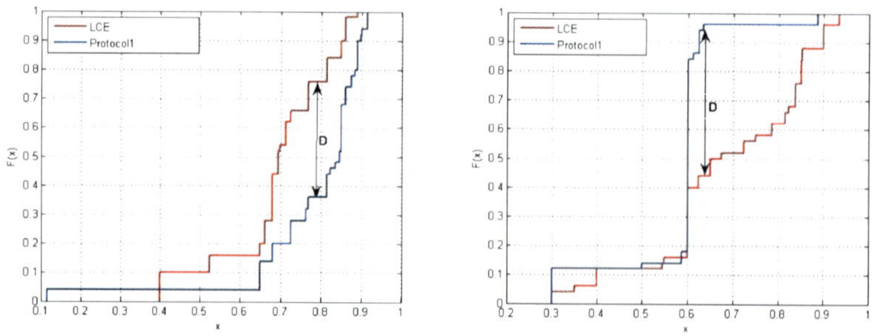

Fig. 4 K-S analysis: the distance between the distributions of utilities obtained in the first concurrent negotiation protocol and the QEA on the buyer side (left) and on the seller side (right)

The above mentioned result is further confirmed by the *Two-Sample Kolmogorov-Smirnov test*, presented below. The *K-S statistics* represents the maximum distance between two one-dimensional and continuous probability distributions. Fig. 4 depicts the results of the K-S analysis. In the case of the buyer, $KS_{calc} = 0.42 > KS_{tab} = 0.2715$ and for the winning seller $KS_{calc} = 0.52 > KS_{tab} = 0.2715$. If we calculate the differences between the practical and the theoretical KS values, we obtain: $Dist_b = 0.1485$ and $Dist_s = 0.2485$. The K-S analysis shows that the two outcome distributions are statistically different, the protocol of Alg. 4 leading to a better outcome for the buyer than the theoretical QEA.

A better outcome for the buyer may be due to the fact that the protocol does not entirely fit the private-value model. The sellers are learning both the buyer's preferences and also those of the competitors'. In fact, as the number of rounds is huge, the protocol approaches a public value setup, each seller learning the profile of his competitor. Thus, as indicated in the literature [16], we can expect the buyer to obtain further positive results than in a private-value case and clearly better than in the second-price auction.

The second possible explanation for these results come from the fact that the sellers are learning agents that estimate the buyer profile. This estimation is not perfect and it often happens that the Bayesian agent to retrieve an offer better than the theoretical QEA outcome. In this case, if the buyer returns back this offer, the ascending property of the protocol does not allow the seller agent to revise it, even if, in the future rounds, it observes that this offer might be improved on his side.

4.2 *The 2nd Negotiation Protocol against QEA*

We tested protocol of Alg. 5 considering a deadline T of 6 minutes (360 seconds) for the negotiation. In 100% of the cases, the winner selected in the negotiation mechanism coincides with that selected in the *Qualitative English Auction*(QEA).

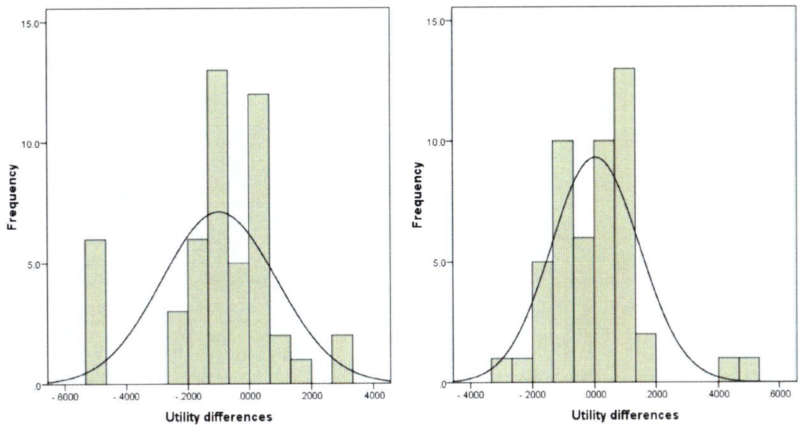

Fig. 5 Utility difference between the second concurrent negotiation protocol and the QEA on the buyer side (left) and on the seller side (right)

Table 4 Synthetic statistical results for utility differences between the first concurrent negotiation protocol and the theoretical QEA

Agent	Mean	Std. Dev.	95% conf. interval Lower	Upper
buyer	-0.097720	0.184711	-0.150214	-0.045226
seller	0.004300	0.141259	-0.035845	0.044445

Fig. 5 presents the distributions of the utility differences between the second negotiation protocol and the QEA. Table 4 presents the 95% confidence intervals for the mean utility values. The average utility difference for the buyer is $\overline{U}^b_{diff} = -0.097720$, which means that in the case of the second negotiation protocol we obtain a slightly lower utility than in the QEA. The confidence interval does not contain 0, which implies that the value is statistically different from the QEA outcome. For the winning seller, the distribution of the utility differences is characterized by a slightly right asymmetry and the average utility difference is $\overline{U}^s_{diff} = 0.004300$. The confidence interval includes 0, which means that he scores obtained almost the same utility as in the theoretical QEA.

Similar results are sustained by the Kolmogorov-Smirnov analysis. Fig. 6 depicts the results of the K-S test. In the case of the buyer, $KS_{calc} = 0.32 > KS_{tab} = 0.2715$ and for the winning seller $KS_{calc} = 0.26 < KS_{tab} = 0.2715$. If we calculate the differences between the practical and the theoretical KS values, we obtain: $Dist_b = 0.0485$ and $Dist_s = 0.0115$.

At this moment, we have the following question: Which concurrent negotiation protocol best approximates the QEA? Based on the calculated distances between the cumulative distributions of the practical and theoretical utilities, we drew the following observations:

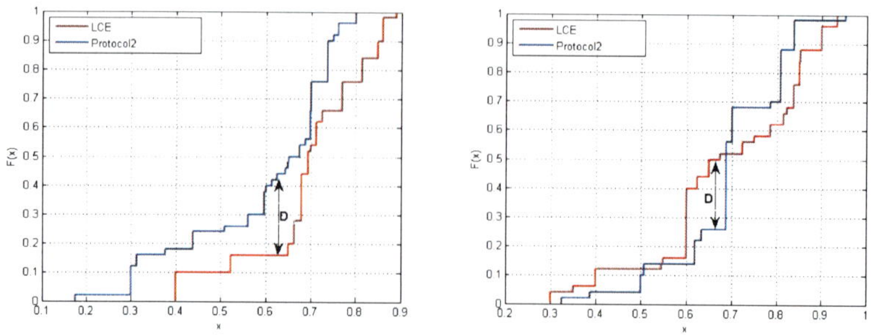

Fig. 6 K-S analysis: the distance between the distributions of utilities obtained in the second concurrent negotiation protocol and the QEA on the buyer side (left) and on the seller side (right)

- *For the buyer*: The concurrent negotiation protocol with deadline approximates better the QEA, because the deviation from the Kolmogorov distribution is only 0.0485, compared to the first protocol (with the withdraw condition), where the deviation is higher (0.1485);
- *For the winning seller*: We obtain the same results, for the second protocol the deviation is 0.0115 compared to the first protocol, where the deviation is 0.2485.

Both protocols leads outcomes on the Pareto frontier. The first protocol, due to the public value model induced by the learning, leads to higher outcome for the buyer and lower for the seller. From the buyer point of view, if his goal is only to increase his utility, this is beneficial. But, because the protocol is long lasting, it lacks practicality, being almost impossible to be utilized in practical scenarios. Further, the first protocol spoils the sellers. Thus, sellers might not have incentives to participate in such an auction, only if the service contract under discussion is of a crucial interest for them (like a high value public bidding).

The second protocol introduces more fairness and practicality. And, because it resembles more with the private value auction, it better approximates the theoretical QEA.

5 Related Work

The concurrent protocol model proposed in this article is based on one-to-many negotiations and it simulates an ascending English auction mechanism. In this section we present other existent models in the literature for trading services characterized by several attributes.

Bilateral negotiation is regarded as the building block for the one-to-many negotiations considered in our setup [11]. The coordination of interactions between participants in such negotiation models is difficult and therefore more structured models, like auctions, were used to simulate them [14]. Various types of direct and reverse auctions were analyzed such as: English [10], Dutch or Vickrey [9]. Auction models are regarded as flexible and provide easy implementations, especially for online applications. In general, they satisfy some market requirements for certain scenarios, but fail in other situations where complex interactions between participants appear, requiring unstructured negotiations. In contrast with a negotiation, an auction differs by the lack of a bidirectional flow of information, the opponent (the auctioneer) being unable to respond with a counter-offer. Also there is no flexible way to practice different negotiation strategies with different opponents on different threads.

One-to-many negotiations can be executed sequentially or concurrently. Running sequential negotiation with each agent individually might be time consuming and is not suitable for setup where real-time service transactions are necessary. Theory proved that the two models do not differ in terms of the utility obtained from participants after negotiation [4].

In the case of concurrent negotiations, difficulties are related to the coordination and the degree of participants' commitment. Nguyen et al. [13] presents a heuristic model for coordinating concurrent negotiations which includes a commitment element for participants, who have now the opportunity to make decisions about continuing or leaving the negotiation. The service providers may break commitment at any time, paying a penalty, but they have the possibility to resume the negotiation later. The negotiation can be seen as a distributed constraint satisfaction problem, where the buyer coordinates the negotiations with various participants by adopting different strategies on each negotiation thread [15]. In our work, we did not attempt to coordinate negotiations with some individual participants; we let each participant to bid actively in every round and give them some hints to make the auction ascending.

6 Conclusions

In this paper we propose two negotiation protocols to approximate the qualitative English auction for multi-issue auction setups with non-transferable utilities. The QVA is proved to possess nice theoretical properties, generalizing the classical Vickrey auction to environments without an established currency. By extension, the same relationship exists between QEA and classical English auctions [6]. The main reason was to remove the major drawback regarding *information revelation*. We show that an optimal outcome can be obtained without the need that the center (auctioneer) to publicly announce her preferences at the beginning of the auction. For each of these mechanisms, the bidders adopt a Bayesian learning strategy for estimating the buyer's preferences, in order to obtain Pareto efficient outcomes. The first negotiation protocol considers the withdrawal of all the bidders but one, as an ending condition, while the second protocol adds a deadline for the negotiation. In the case of the first protocol, the experiments show that this produces outcomes that are still Pareto-efficient, however with a better off buyer and worse off winning seller than in the case of theoretical QEA. This may be due to the fact that the protocol does not anymore fit the private-value model, because sellers learn their competitors profiles. In the case of the second protocol, because of the deadline condition we used a modified version of the Bayesian learning strategy, performing concession by the time elapsed in the negotiation. The experimental results show that the utility differences for the buyer and the winning seller are lower than in the case of the first protocol, which means that this negotiation protocol approximates better the QEA.

Acknowledgements. We acknowledge support from the Romanian Authority for Scientific Research under project IDEI_2452. G.C. Silaghi thanks support from the Sectoral Operational Programme on Human Resources Development under project CommScie POSDRU/89/1.5/S/63663.

References

1. Andrieux, A., Czajkowski, K., Dan, A., Keahey, K., Ludwig, H., Nakata, T., Pruyne, J., Rofrano, J., Tuecke, S., Xu, M.: Web services agreement specification (ws-agreement). In: OGF Grid Resource Allocation Agreement Protocol (GRAAP) WG (2007), `http://www.ggf.org/documents/GFD.107.pdf`
2. Baarslag, T., Hindriks, K., Jonker, C., Kraus, S., Lin, R.: The First Automated Negotiating Agents Competition (ANAC 2010). In: Ito, T., Zhang, M., Robu, V., Fatima, S., Matsuo, T. (eds.) New Trends in Agent-Based Complex Automated Negotiations. SCI, vol. 383, pp. 113–135. Springer, Heidelberg (2012)
3. Bichler, M., Lee, J., Lee, H.S., Chung, J.Y.: ABSolute: An Intelligent Decision Making Framework for E-Sourcing. In: International Workshop on Advanced Issues of E-Commerce and Web-Based Information Systems, pp. 195–201. IEEE Computer Society (2001)
4. Binmore, K., Shaked, A., Sutton, J.: An outside option experiment. The Quarterly Journal of Economics 104(4), 753–770 (1989)
5. Faratin, P., Sierra, C., Jennings, N.: Using similarity criteria to make issue trade-offs in automated negotiations. Artificial Intelligence 142(2), 205–237 (2002)
6. Harrenstein, B.P., de Weerdt, M.M., Conitzer, V.: A qualitative vickrey auction. In: Proceedings of the 10th ACM Conference on Electronic Commerce, pp. 197–206. ACM (2009)
7. Hindriks, K., Jonker, C.M., Kraus, S., Lin, R., Tykhonov, D.: Genius: negotiation environment for heterogeneous agents. In: AAMAS 2009: Proceedings of The 8th International Conference on Autonomous Agents and Multiagent Systems, pp. 1397–1398. IFAAMAS (2009)
8. Hindriks, K., Tykhonov, D.: Opponent modelling in automated multi-issue negotiation using bayesian learning. In: AAMAS 2008: Proceedings of the 7th International Joint Conference on Autonomous Agents and Multiagent Systems, pp. 331–338. IFAAMAS (2008)
9. Hindriks, K., Tykhonov, D., de Weerdt, M.: Qualitative One-to-Many Multi-Issue Negotiation: Approximating the QVA. Group Decision and Negotiation 21(1), 49–77 (2010)
10. Li, C., Giampapa, J., Sycara, K.: Bilateral negotiation decisions with uncertain dynamic outside options. IEEE Transactions on Systems, Man, and Cybernetics, Part C: Applications and Reviews 36(1), 31–44 (2006)
11. Lomuscio, A.R., Wooldridge, M., Jennings, N.R.: A classification scheme for negotiation in electronic commerce. Group Decision and Negotiation 12(1), 31–56 (2003)

12. Milgrom, P., Webber, R.: A theory of auctions and competitive bidding. Econometrica 50(5), 1089–1122 (1982)
13. Nguyen, T.D., Jennings, N.R.: Reasoning about commitments in multiple concurrent negotiations. In: Proceedings of the 6th International Conference on Electronic Commerce, ICEC 2004, pp. 77–84. ACM (2004)
14. Parkes, D.C., Ungar, L.H., Foster, D.P.: Accounting for Cognitive Costs in On-Line Auction Design. In: Noriega, P., Sierra, C. (eds.) AMET 1998 and AMEC 1998. LNCS (LNAI), vol. 1571, pp. 25–40. Springer, Heidelberg (1999)
15. Rahwan, I., Kowalczyk, R., Pham, H.H.: Intelligent agents for automated one-to-many e-commerce negotiation. Australian Computer Science Communications 24(1), 197–204 (2002)
16. Sandholm, T.W.: Distributed rational decision making, pp. 201–258. MIT Press (1999)
17. Silaghi, G.C., Serban, L.D., Litan, C.M.: A Time-Constrained SLA Negotiation Strategy in Competitive Computational Grids. Future Generation Computer Systems (2011) (Online first)
18. Waeldrich, O., Battre, D., Brazier, F., Clark, K., Oey, M., Papaspyrou, A., Wieder, P., Ziegler, W.: Ws-agreement negotiation 1.0. In: OGF Grid Resource Allocation Agreement Protocol (GRAAP) WG (2010)

Iterative, Incremental and Evolving EAF-Based Negotiation Process

Paulo Maio, Nuno Silva, and José Cardoso

1 Introduction

Internally agents may use argumentation for both (i) reasoning about what to believe (i.e. theoretical reasoning) and/or (ii) for deciding what to do (i.e. practical reasoning). Despite existing differences between both, from a standpoint of first-personal reflection, a set of considerations for and against a particular conclusion are drawn on both [1]. On the other hand, concerning the types of agents' dialogues (e.g. Deliberation, Negotiation, Persuasion, Inquiry, Information-seeking dialogues), while a clear distinction between each one exist, most of the agents' dialogue occurrences involve mixtures of dialogue types. Within this context, argumentation is seen as an activity where each participant tries to increase (or decrease) the acceptability of a given standpoint for the others participants by presenting arguments. In particular, agents participating in a negotiation dialogue may use argumentation to support their position and by that achieve a better agreement. Therefore, argumentation is foreseen as an adequate modeling formalism to reduce the gap between models governing the internal and external agent behavior. Grounded on that, this paper presents a novel, generic and domain independent argument-based negotiation process, in which are advocated the benefits of defining a common argumentation vocabulary shared by all agents participating in negotiation, which is internally extended by each of the agents to fit its own needs and knowledge. For that, the proposed argument-based negotiation process also requires that argumentation frameworks have modeling, modular and extensibility features. In that sense, the adoption of the Extensible Argumentation Framework (EAF) is suggested.

Paulo Maio · Nuno Silva
GECAD, School of Engineering, Polytechnic of Porto, Portugal
e-mail: {pam,nps}@isep.ipp.pt

José Cardoso
University of Trás os Montes and Alto Douro, Vila Real, Portugal
e-mail: jcardoso@utad.pt

T. Ito et al. (Eds.): Complex Automated Negotiations, SCI 435, pp. 161–179.
springerlink.com

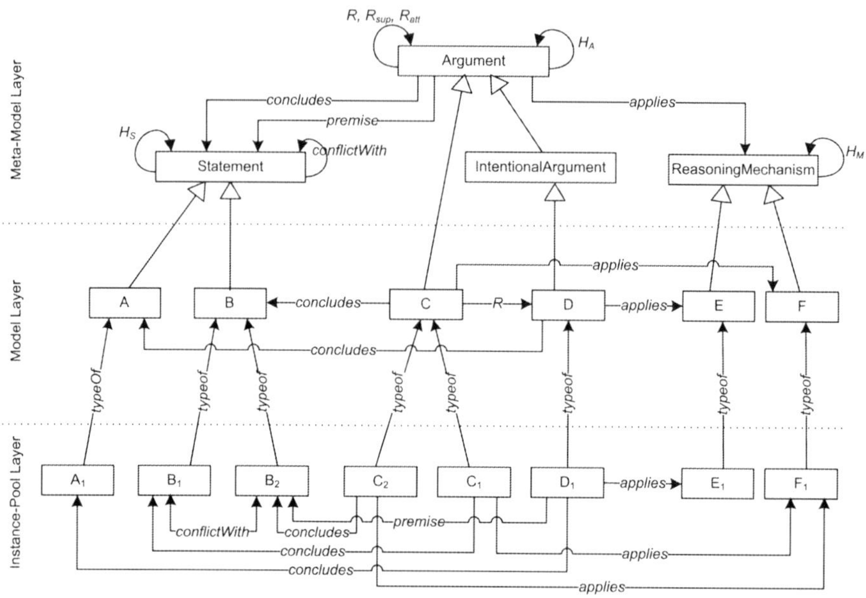

Fig. 1 The three modeling layers of TLAF/EAF

The rest of this paper is organized as follows: the next section describes the main structures and concepts of the EAF. Section 3 presents the proposed negotiation process based on the adoption of EAF in MAS [9]. Section 4 describes and summarizes the performed experiments in the domain of ontology alignment [2] applying the proposed negotiation process. Finally, section 5 draws conclusions and comments on future work.

2 Extensible Argumentation Framework

The Extensible Argumentation Framework (EAF) [10] is based and extends the Three-layer Argumentation Framework (TLAF) [3]. TLAF is a generic argumentation framework that, unlike others (e.g. AF [4], BAF [5], VAF [6]) comprehends three modeling layers. While the Meta-Model Layer and the Instance Layer roughly correspond to the (meta-) model layer and the instance layer of AF, BAF and VAF, the Model Layer does not have any correspondence in the surveyed abstract argumentation frameworks (illustrated in Fig. 1).

The Meta-model layer defines the core argumentation concepts and relations holding between them. TLAF adopts and extends the minimal definition presented by Walton in [7] where "an argument is a set of statements (propositions), made up of three parts, a conclusion, a set of premises, and an inference from premises to the conclusion". For that, the meta-model layer defines the notion of *Argument*,

Statement and *Reasoning Mechanism*, and a set of relations between these concepts. An argument *applies* a reasoning mechanism (such as rules, methods, or processes) to *conclude* a conclusion-statement from a set of premise-statements. Intentional arguments are the arguments corresponding to intentions ([8, 9]).

The Model layer represents a conceptualization of arguments by defining the entities and their relations regarding both:

- the domain of application to be captured;
- the perception one (e.g. a community of agents or an individual agent) has about that domain.

The resulting model is further instantiated at the Instance-pool layer. The R relation is established between two argument types (e.g. $(C,D) \in R$) when C supports or attacks D. Through R it is also determined the types of statements that are admissible as premises of an argument.

Fig. 2 partially and graphically represents a simple argumentation model on the ontology matching domain. In this model, the intention to accept/reject a given correspondence between two ontological entities is captured by an argument of type *MatchArg* concluding a statement of type *MatchSt*. An argument of type *MatchArg* is affected (either supported or attacked) by arguments of type *TerminologicalArg* and *StructuralArg* concluding statements of type *TerminologicalSt* and *StructuralSt* respectively. All these arguments apply an *Heuristic* reasoning mechanism (not depicted in Fig. 2).

The Instance-Pool layer corresponds to the instantiation of a particular model layer for a given scenario. A statement-instance B_1 is said to be in conflict with another statement-instance B_2 when B_1 states something that implies or suggests that B_2 is not true. The statement conflict relation is asymmetric (in Fig. 1 B_2 conflicts with B_1 too). The support and attack relationships (R_{sup} and R_{att} respectively) between argument-instances are automatically inferred exploiting:

- the conceptual information (existing at the model layer), namely the R relations defined between argument-types;
- the extensional information (existing at the instance layer):
- the premises and conclusions of the argument-instances;
- the conflicts between statement-instances.

EAF extends TLAF by providing the constructs and respective semantics for supporting modularization and extensibility features to TLAF. In that sense, any EAF model is a TLAF model but not the inverse. In EAF model layer, arguments, statements and reasoning mechanisms can be structured through the H_A, H_S and H_M relations respectively. These are acyclic transitive relations established between similar entity types (e.g. arguments), in the sense that in some specific context entities of type e_1 are understood as entities of type e_2. While these relations are vaguely similar to the specialization relation (i.e. subclass/superclass between entities) it does not have the same semantics and it is constrained to 1-1 relationship (cf. [10]). An EAF model may reuse and further extend the argumentation conceptualizations of several existing EAF models. Inclusion of an EAF into another EAF is governed by

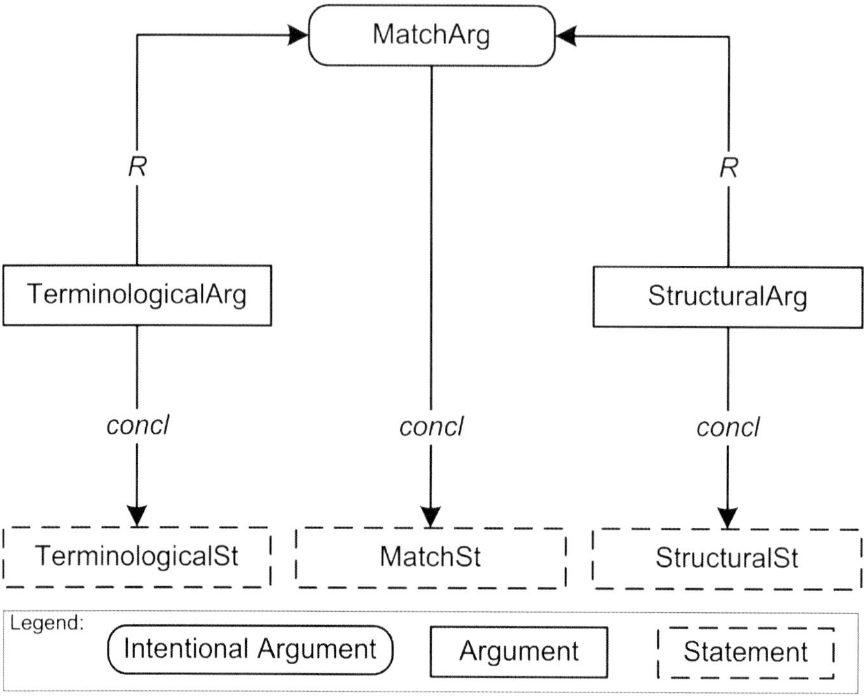

Fig. 2 Example of a TLAF model for the ontology matching domain

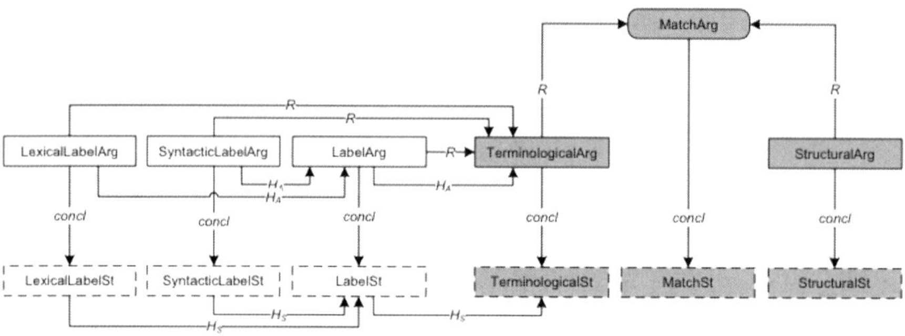

Fig. 3 Example of an EAF model with $\mathbf{H_A}$ and $\mathbf{H_S}$ relations

a set of modularization constraints ensuring that no information of included EAF is lost. The extensibility feature of EAF is illustrated in the example depicted in Fig. 3 regarding the ontology matching domain.

The EAF model depicted in this figure (called EAF_{OM_1}) extends the TLAF/EAF model previously depicted in Fig. 2 (called EAF_{OM}) such that the new arguments and statements are colored white while the arguments and statements of the extended model are colored gray. According to this example, the EAF semantics imply (for example) that any instance of *LexicalLabelArg* is understood and is translatable to an instance of *LabelArg*, which in turn is translatable into an instance of *TerminologicalArg*. In the argument exchange context, this feature is relevant considering that each agent internally adopting a distinct EAF model (e.g. EAF_{OM_1}) extended from a common/shared EAF model (e.g. EAF_{OM}) may translate arguments represented in their internal model to the shared model and, therefore, enabling the understanding of those arguments by the other agents.

3 The Argument-Based Negotiation Process

This section proposes the argument-based negotiation process (ANP) based on the adoption of EAF by agents in MAS [9]. While other negotiation processes using EAF are admissible, we aim to provide an end-to-end negotiation process that emphasizes its potential and applicability.

3.1 Principles

Observations show that agreements through argumentation between humans follow an iterative and incremental process where arguments and counter-arguments are successively presented, enabling humans to identify the existing conflicts and further present more arguments and counter-arguments to (tentatively) resolve such conflicts.

Concerning the arguments formulation, humans usually exploit a huge diversity of information sources which may provide information that is more or less reliable, (in)complete, (in)coherent, (in)consistent and so on. Thus, each human usually selects and exploits information provided by the sources that are considered more reliable and trustable for the problem in hands.

Concerning the arguments understanding and reasoning, each human has a unique (i.e. its own) perception and rationality over the domain of the problem in hands. Therefore, arguments are seen, interpreted and evaluated in light of that individual perception. This fact enables humans to extract from the same set of arguments several distinct and contradictory conclusions.

Typically, the argumentation process ends either (i) when no more conflicts exist or (ii) when no more arguments are presented by any of the participants. In the former case, the argumentation always ends successfully since no conflicts exist anymore. In the latter case, the argumentation may end successfully or unsuccessfully depending on the degree of importance that each one gives to the remaining conflicts when compared to the agreement in hand. Thus, if the parties agree that the agreement in hand is better than no agreement at all then the argumentation ends successfully, otherwise it ends unsuccessfully.

At least but not less important, with respect to human beings' natural ability to evolve their knowledge and perception of the world and particularly about the domain under which they are arguing. A classical situation occurs when a human faces an argument put forward by another human and (s)he does not know its meaning or how that argument relates and affects the others known arguments. In such cases, that human may require a conceptual description of that kind of argument in order to figure out the missing knowledge and therefore acquire it. The resulting knowledge acquisition contributes to the evolution of its perception of the domain under discussion.

3.2 *Overview*

Within the proposed ANP the negotiation entities (e.g. persons, organizations) are represented by agents. Yet, as with any other negotiation process, the proposed argument-based negotiation process happens, at least, between two agents. Furthermore, it is assumed that the negotiation occurs in the scope of a given community of agents. When joining a community, the agent is (implicitly or explicitly) accepting a set of rules by which all agents interactions are governed. One of the main rules is related to the key notion/concept of argumentation model, which in turn substantially constrains the characteristics of the argumentation process.

Definition 1 (Argumentation Model). An argumentation model (AM) is an artifact that captures (partially or totally) the perception and rationality that an agent has about a specific domain (e.g. ontology matching) regarding the argumentation process.

According to Definition 1, the argumentation model might conceptually define the vocabulary used to form arguments, the arguments' structure and even the way arguments affect (i.e. attack and support) each other. Hence, a model is a specification used for stating model commitments. In practice, a model commitment is an agreement to use a vocabulary in a way that is consistent (but not necessarily complete) with respect to the theory specified by the model [11]. Agents commit to models which are designed so that the domain knowledge can be shared among these agents.

The community of agents on which the negotiation process occurs is responsible for defining a public argumentation model.

Definition 2 (Public Argumentation Model). A public argumentation model is a shared argumentation model capturing the common understanding about argumentation over the domain problem being addressed (e.g. ontology matching) of a community of agents.

All agents of that community are able to understand the defined public argumentation model and reason on it. Further, each agent must be able to extend the public argumentation model so it better fits its own needs and knowledge. As a result, the agents freely specify their private argumentation model.

Definition 3 (Private Argumentation Model). A private argumentation model is an argumentation model capturing the understanding about argumentation over the domain problem being addressed (e.g. ontology matching) of a single agent.

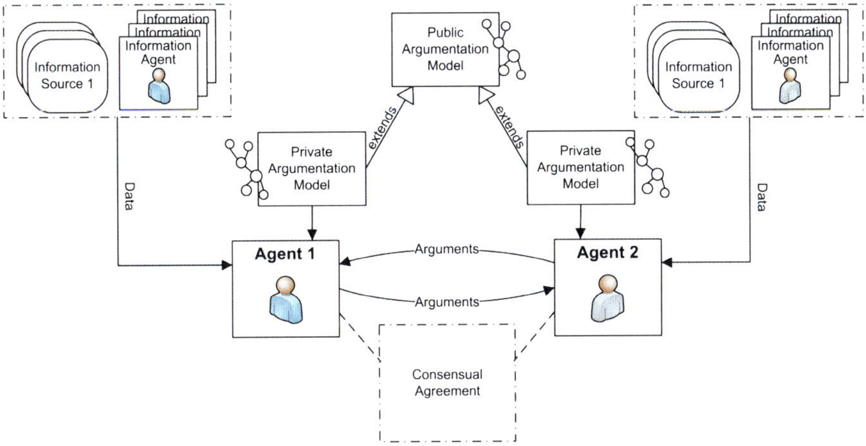

Fig. 4 Overview of the argument-based negotiation process

While a public argumentation model represents a shared knowledge/perception between agents, a private argumentation model represents the individual perception/knowledge that an agent has.

Because the agents adopt their own private argumentation model, each agent has the responsibility for searching, identifying and selecting the sources of information that can provide the most relevant and significant information needed to instantiate its private model. After the private model instantiation each agent has a set of arguments that need to be evaluated in order to extract the agent consistent position, i.e. a *preferred extension*. A *preferred extension* includes two kinds of argument: intentional arguments and non-intentional arguments. The former ones define the intentions of the agent with respect to the agreement, while the latter ones represent the set of reasons supporting the intentions. By exchanging the intentional arguments of their *preferred extensions*, agents are able to identify the existing conflicts and argue with non-intentional arguments.

It is worth noticing that the EAF model layer together with the extensibility and modularization features satisfies the above definitions of public/private argumentation model. Therefore, from now the ANP description adopts EAF.

3.3 Phases of the Negotiation Process

Considering the premises described in previous section, a general argument-based negotiation process was devised. The phases of each agent's negotiation process, the flow of data and the interactions with other agents are depicted in Fig. 5.

3.3.1 Setup

The Setup phase defines the context of the negotiation. At the end of this phase all participating agents know and agree with this context. For that, the participating agents will engage in a set of interactions aiming for:

- the identification of the (possible) negotiation participants;
- the identification of the community's minimal common understanding, i.e. the public argumentation model (EAF_C) between all participants;
- the definition of the required negotiation parameters/constraints such as deadline for achieving an agreement;
- the specification of the negotiation method to compute a possible agreement between participants;
- the establishment of special rights for some of the participants;
- the sharing of the data/information that is required by the agents in order to participate in the negotiation (e.g. the ontology used by each agent).

These interactions will result in the definition of a set of constraints called the negotiation parameters (NP). Complementary to the negotiation parameters, each participant creates an instance-pool of its own argumentation model ($IP(EAF_{Ag})$) that will capture the argumentation data. In contrast to the other phases, this phase occurs only once.

3.3.2 Data Acquisition

During the Data Acquisition phase the agent collects the data/information that constitutes the grounds to generate the arguments (called D_{Ag}). For that, the agents may interact with other agents not directly participating in the negotiation process. It might be the case of specialized agents on the subject under discussion. The tentative agreements generated in the upcoming phases may be used as input information to the data-collecting mechanisms too.

3.3.3 Argument Instantiation

The goal of the Argument Instantiation phase is the instantiation of the agent's instance-pool of the argumentation model ($IP(EAF_{Ag})$) based on the collected data (D_{Ag}). For that, the agent makes use of one or more data transformation processes over the collected data, generating a set of arguments structured according to the adopted argumentation model. In order to properly (re)classify the argument instances is foreseen the need of an instances (re)classification process. It is also envisaged that this (re)classification process might be further reused in the Instance-Pool Update phase. However that is not mandatory. This process is extensively addressed in [3] and [12].

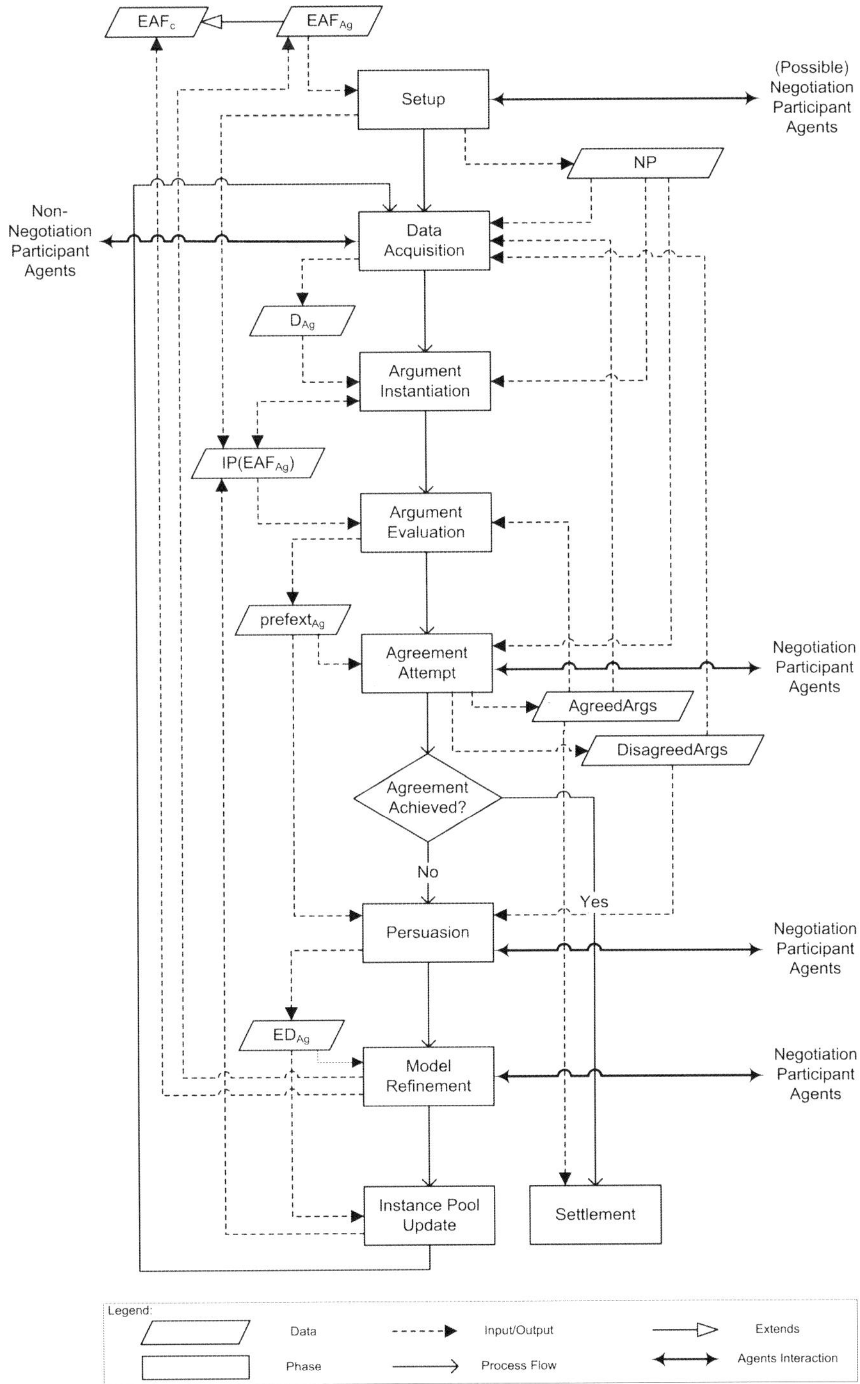

Fig. 5 The proposed argument-based negotiation process

3.3.4 Argument Evaluation

In the Argument Evaluation phase, each agent extracts a *preferred extension*, i.e. a consistent position within $IP(EAF_{Ag})$ which is defensible against any attack and cannot be further extended without introducing a conflict. According to the agent's $IP(AM_{Ag})$ one or more possible *preferred extensions* may be extracted.

If the argument evaluation process extracted more than one *preferred extension* then it is necessary to select one. The selection criterion has a special relevance during the negotiation process because it directly defines the agent's intentions and the reasons behind those intentions. Given that, instead of a simple criterion, a more elaborate selection criterion may be taken into consideration. For example, instead of the "selection of the *preferred extension* that is maximal with respect to set inclusion", one may consider "the *preferred extension* that minimizes the changes in respect to the previous one.

This phase occurs iteratively depending on the acquisition of new data/information and especially on the exchange of arguments between the agents during the persuasion phase. Because any change made to $IP(EAF_{Ag})$ suggests that the agent's consistent position may change, a re-evaluation of the *preferred extension* is necessary.

3.3.5 Agreement Attempt

In the Agreement Attempt phase each participant makes an agreement proposal to the other agent(s) (called the candidate agreement). If accepted it will be settled by all participants.

This phase consists of two steps. In the first step, each agent makes its agreement proposal by exchanging the intentional arguments of its *preferred extension* only (called the *intentional preferred extension*). As a result of all proposals, two sets of arguments are derived and shared by all agents:

- the set of arguments agreed/proposed by all agents ($AgreedArgs$) which represents a candidate agreement;
- the set of arguments which at least one agent disagrees ($DisagreedArgs$). For a negotiation between n agents where $iprefext_{Ag_i}$ is the *intentional preferred extension* of agent i, these sets can be computed differently depending on the agents and according to the setup phase. One of the simplest agreement evaluation forms is based on their intersection:

$$AgreedArgs = \bigcap_{i=1}^{n} iprefext_{Ag_i}$$

$$DisagreedArgs = \left(\bigcup_{i=1}^{n} iprefext_{Ag_i}\right) - AgreedArgs$$

In the second step, each participant evaluates its level of satisfaction of the current candidate agreement. For that the agent considers the defined negotiation parameters/constraints (*NP*) and the content of the *DisagreedArgs* set. According to the level of satisfaction, the participants must decide to whether to:

- continue the negotiation, and therefore proceed to the persuasion phase; or
- conclude the negotiation, which is either:
- successful if all agents accept the candidate agreement (*AgreedArgs*). In such case the process proceeds to the settlement phase; or
- unsuccessful if the candidate agreement is not accepted by all agents and they do not continue the negotiation. In this case no agreement is achieved.

3.3.6 Persuasion

In the previous phase a set of conflicts/disagreements have been identified (in the form of intentional arguments) that were not accepted by at least one participant (*DisagreedArgs*). In this phase each agent tries to persuade the others to accept its intentions. For that, each agent exchange arguments supporting its *preferred extension* and arguments attacking the other agents' *preferred extension(s)*.

Each agent first selects from its *preferred extension* a (sub-) set of arguments supporting or attacking the arguments existing in *DisagreedArgs*. The selected arguments will be exchanged with the opponent agents to persuade them. There are two forms of exchanging the arguments:

i. The arguments are exchanged according to the EAF_c and not according to EAF_{Ag}, so the other agents can understand them. Due to the H_A, H_S and H_M relations, the transformation of the instances respecting the agent's EAF to the community EAF is straightforward. Thus, arguments represented according to the agent's argumentation model EAF_{Ag} that cannot be expressed in terms of EAF_C are not exchanged;
ii. The arguments are exchanged according to the EAF_{Ag} along with the EAF_{Ag} parts that allow the other agent to transform the arguments to EAF_c. This arguments exchange method requires that agents have the ability to teach and to learn from other agents such that agents may evolve over time their perception/knowledge.

Independently of the exchanged method (decided in the Setup phase), at the end of this phase, each agent has collected a new set of arguments presented by the other negotiating agents (ED_{Ag}). These arguments will be exploited in the Instance-Pool Update phase.

3.3.7 Model Refinement

This phase concerns the refinement of the community's argumentation model (EAF_C) according to the exchanged arguments and the agents' argumentation models. If the exchange of arguments does not include exchanging parts of the agent's private argumentation model, this phase is more difficult and therefore may be skipped.

While it is not the aim of this description to present an evolution process of the argumentation model, nor the agents' reasoning process leading to such evolution, it is important to emphasize the need to evolve (over time) the community's argumentation model according to the agents' needs.

3.3.8 Instance-Pool Update

In this phase, the agent analyzes, processes and possibly reclassifies the ED_{Ag} arguments in light of its EAF_{Ag}. The ED_{Ag} arguments that are understood (in the light of EAF_{Ag} or EAF_C) and do not exist in $IP(EAF_{Ag})$ are added while duplicated arguments are discarded. The added arguments are taken into consideration by the agent in the next round of proposals. The negotiation process proceeds to the Data Acquisition phase.

3.3.9 Settlement

The goal of the settlement phase is to transform the candidate agreement into a definitive agreement according to the settlement parameters of *NP*. In that respect, this phase is seen as an initiator of a set of transactions that occur after the agreed terms are known in order to fulfill the terms.

The set of transactions varies according to the domain of application and the negotiation object (e.g. goods or services) as well as the participating agent. On the other hand, in an e-commerce scenario, fulfilling an agreement for selling physical goods may imply forwarding the agreement to the logistic and financial services.

4 Experiments

Since the proposed negotiation approach is domain independent, one needs a domain of application to evaluate the proposed argument-based negotiation approach. We choose to resolve conflicts arising between agents when they are reconciling the vocabulary used in their ontologies. The result of the vocabulary reconciliation is a set of correspondences (an alignment) between entities of the agents' ontologies. Such conflicts arise because each agent may have its own perspective about what are the best correspondences. The experiments aim to measure the (i) conflicts resolved and (ii) the improvement produced in the accuracy of the agreed alignment, when compared to each agent's initial alignment.

4.1 Setup

We adopted an empirical approach using:

- a publicly available set of pairs of ontologies exploited in several ontology alignment initiatives (Table 1);

Table 1 Set of pair of ontologies used in the experiments

Source Ontology	Target Ontology	Nr. Correspondences
animalsA[1]	animalsB[1]	24
russia1[1]	russia2[1]	65
russiaA[1]	russiaB[1]	160
russiaC[1]	russiaD[1]	135
sportEvent[1]	sportSoccer[1]	149
Vehicles1[1]	Vehicles2[1]	4
onto101[2]	onto1032	97
onto101[2]	onto104[2]	97
onto101[2]	onto204[2]	97
onto101[2]	onto205[2]	97
onto101[2]	onto221[2]	97
onto101[2]	onto222[2]	93
onto101[2]	onto223[2]	97
onto101[2]	onto301[2]	45
onto101[2]	onto302[2]	37
onto101[2]	onto303[2]	32
onto101[2]	onto304[2]	56

- for each pair of ontologies a widely accepted reference alignment that will be used to evaluate the negotiation results. For the sake of brevity and simplicity, the results are presented considering the negotiation of all individual alignments as just one huge alignment with 1402 correspondences;
- three agents (A, B and C) with different configurations, including:
- different EAF models (EAF_{Ag_A}, EAF_{Ag_B} and EAF_{Ag_C}) extended from a common one. These EAF models are illustrated in Fig. 6, Fig. 7 and Fig. 8 respectively, where grayed arguments refer to the community's common argumentation model;
- distinct set of matching algorithms (or matchers) were used to collect ontological correspondences which are further used to generate argument-instances (Table 2, Table 3 and Table 4);
- based on the H_A, H_S and H_M, the agents have the capability to reclassified internally the argument-instances as depicted in Table 5. Notice that this table reflects the agents' internal and private knowledge, thus an agent does not know the reclassification rules of the other agents.

The adopted argument instantiation process is described in [12]. It relies on an interpretation function that allows transforming a correspondence $c = (e, e', r, n)$ provided by a matching algorithm (G) into a statement-instance $s = (G, c, pos)$ such that $pos \in \{+, -\}$ states if the matcher is in favor ($+$) or against ($-$) the correspondence c. The matcher position is determined based on the thresholds tr_+ and

[1] Available at http://www.dei.isep.ipp.pt/~pmaio/goals/Ontologies/

[2] Available at http://oaei.ontologymatching.org/2009/benchmarks/

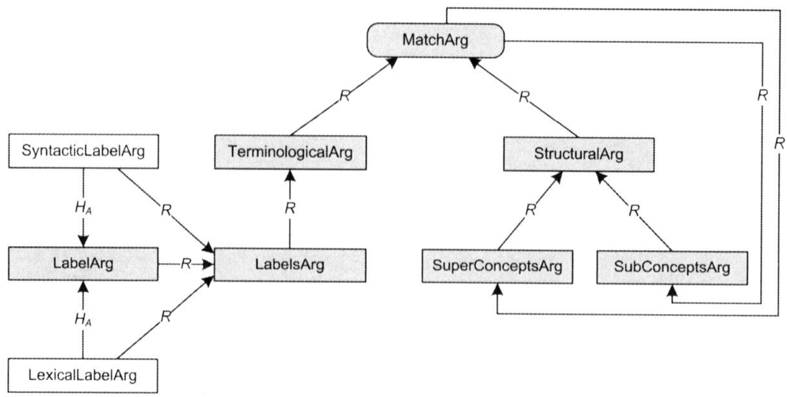

Fig. 6 The argumentation model internally adopted by Agent A

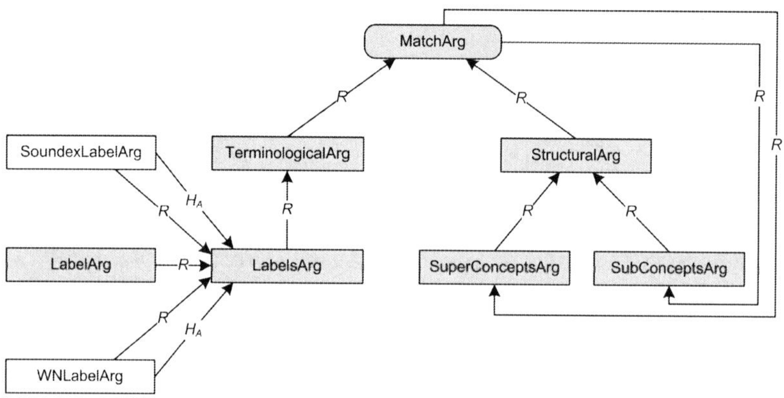

Fig. 7 The argumentation model internally adopted by Agent B

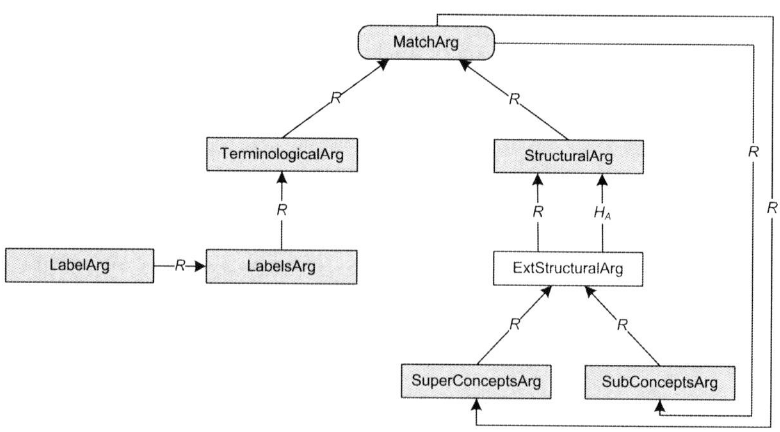

Fig. 8 The argumentation model internally adopted by Agent C

Table 2 The interpretation function of Agent A

ID	Matcher Description	Correspondence Content	Statement Type	Reasoning Mechanism	tr_+	tr_-
G_{A1}	WNMatcher [13]	any	LexicalLabelSt	Heuristic	1.00	1.00
G_{A2}	String-distance [14]	any	SyntaticalLabelSt	Heuristic	0.75	0.75
G_{A3}	V-Doc [14]	any	LabelSt	Heuristic	0.70	0.70
G_{A4}	$MaxAgg(G_{A1}, G_{A2})$[3]	any	TerminologicalSt	Heuristic	0.50	0.50
G_{A5}	GMO [14]	any	SuperConceptsSt	Heuristic	0.50	0.50
G_{A6}	Falcon-AO [14]	any	MatchSt	Heuristic	0.70	0.70

Table 3 The interpretation function of Agent B

ID	Matcher Description	Correspondence Content	Statement Type	Reasoning Mechanism	tr_+	tr_-
G_{B1}	Soundex [15][4]	any	SoundexLabelSt	Heuristic	0.75	0.75
G_{B2}	WNPlusMatcher [13]	any	WNLabelSt	Heuristic	1.00	1.00
G_{B3}	BiGram[5]	any	LabelSt	Heuristic	0.75	0.75
G_{B4}	$OWAAgg(G_{B1}, G_{B2}, G_{B3})$[6]	any	TerminologicalSt	Heuristic	0.60	0.60
G_{B5}	StructureMatcher [13]	any	SuperConceptsSt	Heuristic	0.70	0.70
G_{B6}	Sub-hierarchy [18]	any	SubConceptsSt	Heuristic	0.30	0.30
G_{B7}	$MaxAgg(G_{B2}, SMOA$ [19])	any	MatchSt	Heuristic	0.25	0.25

Table 4 The interpretation function of Agent C

ID	Matcher Description	Correspondence Content	Statement Type	Reasoning Mechanism	tr_+	tr_-
G_{C1}	Levenshtein [20]	any	LabelSt	Heuristic	1.00	1.00
G_{C2}	WNPlusMatcher [13]	any	LabelSt	Heuristic	0.75	0.75
G_{C3}	$AvgAgg(G_{C1}, G_{C2}, SMOA)$[7]	any	TerminologicalSt	Heuristic	0.70	0.70
G_{C4}	Super-hierarchy [18]	any	SuperConceptsSt	Heuristic	0.70	0.70
G_{C5}	Sub-hierarchy [18]	any	SubConceptsSt	Heuristic	0.70	0.70
G_{C6}	$AvgAgg(G_{B5}, SMOA)$	any	StructuralSt	Heuristic	0.70	0.70
G_{C7}	$Op(MaxAgg(G_{C2}, SMOA, G_{B5})$[8]	any	MatchSt	Heuristic	0.25	0.25

tr_- such that: if $n \geq tr_+$ then the matcher is in favor of c, otherwise if $n < tr_-$ then the matcher is against c. If the matcher is neither for nor against c the correspondence is ignored. The resulting statement-instance is further concluded by an argument-instance based on a set of inferred premise statement-instances. The interpretation functions adopted by agent A, B and C are represented in Table 2, Table 3 and Table 4 respectively.

As an example, argument-instances sent by agent A whose conclusion is a statement-instance of type *LexicalLabelSt* that were generated based on a correspondence provided by G_{A1} are further reclassified by agent B to statement-instances of type *WNLabelSt* and, therefore, to argument-instances of type *WNLabelArg*.

[3] Corresponds to the aggregation of the alignments generated by the input matching algorithms through the *max* function.

[4] Implemented in the SimMetrics project available at http://sourceforge.net/projects/simmetrics/.

[5] Implemented in the SimPack [16]

[6] Corresponds to the aggregation of the alignments generated by the input matching algorithms through the OWA operator[17]

[7] Corresponds to the aggregation of the alignments generated by the input matching algorithms through the linear average function.

[8] Corresponds to the global optimization of the input alignment by the Hungarian method [21].

Table 5 Reclassification of statement-instances based on its content

Agent	Statement-Content		EAF_A		EAF_B
	Type	Matcher	LexicalLabelSt	SyntaticLabelSt	WNLabelSt
A	LexicalLabelSt	G_{A1}	-	-	X
B	WNLabelSt	G_{B2}	X	-	-
B	SoundexLabelSt	G_{B3}	-	X	-
C	LabelSt	G_{C1}	X	-	X
C	LabelSt	G_{C2}	-	X	-

Table 6 Agents' alignment before the negotiation process

Agent	Correspondences		Accuracy (%)		
	Proposed	Correct	Precision	Recall	F-Measure
A	1358	1296	95.4	92.4	93.9
B	2025	1266	62.5	90.3	73.9
C	1290	1219	94.5	86.9	90.6

Table 7 Agreed Alignment between agents

Agents	Correspondences		Accuracy(%)		
	Proposed	Correct	Precision	Recall	F-Measure
A-B	1294	1243	96.1	88.7	92.2
A-C	1250	1214	97.1	86.6	91.6
B-C	1387	1234	89.0	88.0	88.5

4.2 *Results*

Table 6 summarizes and characterizes the automatic alignment of each agent before the negotiation process by presenting the (information retrieval) measures of Precision, Recall and F-Measure[9]. Correct correspondences are those that exist in the reference alignment.

Table 7 summarizes and characterizes the agreed alignment between each possible pair of agents.

Table 8 summarizes and characterizes the amount of conflicts addressed during the negotiation process and the quality of the occurred persuasion. For each pair of agents, the table shows:

- the amount of conflicts before the negotiation process starts and the amount of those conflicts that are about correspondences belonging to the reference alignment (R.A);
- the amount of conflicts that remain to be resolved at the end of the negotiation process and the amount of those conflicts that are about correspondences belonging to the reference alignment (R.A);
- the amount of resolved conflicts and the corresponding amount of those conflicts that were correctly and badly resolved regarding both the agreed alignment and the reference alignment;
- the rate of persuasion occurred between the agents (i.e. rate of resolved conflicts) and the quality of that persuasion.

[9] F-Measure is the harmonic mean of Precision and Recall.

Table 8 Conflicts addressed during the negotiation process

Agents	Conflicts							Persuasion (%)		
	Initial		Remain		Resolved					
	Total	R.A.	Total	R.A.	Total	Correct	Bad	Total	Good	Bad
A-B	813	78	308	67	505	487	18	62.1	96.4	3.6
A-C	200	119	130	90	70	47	23	35.0	67.1	32.9
B-C	779	75	223	39	556	459	97	71.7	82.6	17.4

4.3 *Analysis and Discussion*

Regarding the conflicts resolution, the examination of the results shows that the proposed negotiation process allows agents to resolve their conflicts. However, as it was expected the amount of resolved conflicts depends of the persuasiveness of the negotiating agents. In that sense, several levels of persuasiveness were observed (from 35% to 71.7%). Even though, the rate of good persuasion is always high since it varies between 67.1% and 96.4%. While these results may depend of several and distinct factors, one might conclude that they are independently of the amount of resolved conflicts.

Yet, comparing the initial amount of conflicts about correspondences belonging to the reference alignment and the amount of those conflicts that remains when the negotiation process ends, it is perceivable that it is very hard for an agent to successfully persuade its opponent to change position about a correct correspondence proposed by its opponent.

Comparing the alignment devised individually by the agents with the agreed alignment, one might say that agents profit with the argumentation process:

- agent A agreed two alignments which are, in terms of f-measure, at maximum 2.3% worse than the one devised by itself;
- agent B improved its alignment whose f-measure is around 74% to an agreed alignment whose f-measure is at least 88%, i.e. an increase of 14%;
- agent C agreed an alignment with agent B which is, in terms of f-measure, around 2% worse than the one devised by itself, but it has improved around 1% in the agreed alignment with agent A;

These f-measure variations happened at the same time that conflicts are resolved. Considering that, differences observed in the f-measure of the agreed alignments of agent A and C are negligible. On the other hand, the improvement achieved by agent B is very significant.

5 Conclusions

This paper describes conceptually a novel, generic and domain independent argument-based negotiation occurring between parts. Thus, it does not specify/defend/provide any implementation details, such as (i) the agents' communication language to adopt, (ii) the exchanging messages, their structure and protocol, (iii) the algorithms to be adopted in each task/phase and (iv) the data sources to be exploited. While all these dimensions are important, the abstraction proposed by the ANP allows:

- to identify the core notion of argumentation model and its influence on the other dimensions;
- to systematize and organize the phases of a negotiation based on arguments from the perspective of an agent and, therefore, the phases become task-oriented;
- to clearly identify the actors and their roles in the negotiation process and namely on which phases they act and interact;
- to identify the main blocks of data/information used as input and as output of the phases of the process;

In face of that, the proposed ANP also advocates:

- an iterative and incremental flow of the results between the phases of the negotiation process;
- four characteristics to any argumentation model: sharable, reusable, extensible and modular;
- the idea that the negotiation participants have distinct knowledge and perspectives about the domain application being argued/negotiated;
- that despite the negotiation participants have distinct knowledge and perspectives about the domain application, a part of that knowledge is shared to foster the exchanged argument understanding;

As a consequence, the process is sufficiently generic to be adopted in a wide range of domains.

This paper also suggested the adoption of the Extensible Argumentation Framework because, contrary to the abstract argumentation frameworks such as AF [4], BAF [5], VAF [6], it comprehends a modeling layer whose content is sharable and reusable through its modularity and extensibility features. By adopting the EAF, agents are able to share an external common argumentation model which is further extended internally by each agent to better fit its own needs and knowledge. Yet, the common argumentation model may continuously evolve along the time profiting from occurring negotiation interaction between agents.

The proposed negotiation process also promotes the use of argumentation as a common formalism either for (i) agents' internal reasoning and (ii) agents interactions (namely negotiation interactions).

Experiments in the ontology alignment field show that the adoption of the proposed ANP together with the adoption of the EAF leads to improvements in the quality of the agreed ontology alignment when compared with agents' individual ontology alignment while conflicts are resolved. High rates of good persuasion are achieved independently of the amount of resolved conflicts.

An interesting research direction concerns providing agents with the ability (i) to learn and improve their argumentation strategies based on their past experiences and (ii) to learn (and understand) new arguments used by other agents in order to apply in the Community's Model Refinement phase.

Acknowledgements. This work is partially supported by the Portuguese MCTES-FCT project COALESCE (PTDC/EIA/74417/2006).

References

1. Moran, R.: Authority and Estrangement: An Essay on Self-Knowledge. Princeton University Press (2001)
2. Euzenat, J., Shvaiko, P.: Ontology Matching. Springer, Heidelberg (2007)
3. Maio, P., Silva, N.: A Three-Layer Argumentation Framework. In: First Int. Workshop on the Theory and Applications of Formal Argumentation at IJCAI, Barcelona, Spain (2011)
4. Dung, P.M.: On the acceptability of arguments and its fundamental role in nonmonotonic reasoning, logic programming and n-person games. Artificial Intelligence 77, 321–358 (1995)
5. Cayrol, C., Lagasquie-Schiex, M.C.: On the Acceptability of Arguments in Bipolar Argumentation Frameworks. In: Godo, L. (ed.) ECSQARU 2005. LNCS (LNAI), vol. 3571, pp. 378–389. Springer, Heidelberg (2005)
6. Bench-Capon, T.: Persuasion in Practical Argument Using Value-based Argumentation Frameworks. Journal of Logic and Computation 13, 429–448 (2003)
7. Walton, D.: Argumentation Theory: A Very Short Introduction. In: Argumentation in Artificial Intelligence, Boston, MA, pp. 1–22. Springer, US (2009)
8. Bratman, M.E.: Intention, Plans, and Practical Reason. Center for the Study of Language and Information (1999)
9. Wooldridge, M.: An Introduction to MultiAgent Systems. John Wiley & Sons (2009)
10. Maio, P., Silva, N.: Tech. report: The Extensible Argumentation Framework. ISEP, Porto, Portugal (2011)
11. Gruber, T.R.: A translation approach to portable ontology specifications. Journal of Knowledge Acquisition 5, 199–220 (1993)
12. Maio, P., Silva, N., Cardoso, J.: Generating Arguments for Ontology Matching. In: 10th Int. Workshop on Web Semantics (WebS) at DEXA, Toulouse, France (2011)
13. Kalfoglou, Y., Hu, B., Shadbolt, N., Reynolds, D.: CROSI-Capturing Representing and Operationalising Semantic Integration, `http://www.aktors.org/crosi/`
14. Jian, N., Hu, W., Cheng, G., Qu, Y.: Falcon-AO: Aligning Ontologies with Falcon. In: Proc. K-CAP Workshop on Integrating Ontologies, pp. 87–93 (2005)
15. Russell, R.C.: US Patent 1261167 (A) (1918)
16. Bernstein, A., Kaufmann, E., Kiefer, C., Burki, C.: SimPack: A Generic Java Library for Similarity Measures in Ontologies. Tech. report, Univ. Zurich, Dpt. of Informatics (2005)
17. jI, Q., Haase, P., Qi, G.: Combination of Similarity measures in Ontology Matching using the OWA Operator. In: Proc. of the 12th Int. Conf. on Information Processing and Management of Uncertainty in Knowledge-Base Systems, IPMU 2008 (2008)
18. Maio, P., Silva, N.: GOALS - A test-bed for ontology matching. In: 1st IC3K Int. Conf. on Knowledge Engineering and Ontology Development, KEOD, Funchal, Portugal (2009)
19. Stoilos, G., Stamou, G., Kollias, S.: A String Metric for Ontology Alignment. In: Gil, Y., Motta, E., Benjamins, V.R., Musen, M.A. (eds.) ISWC 2005. LNCS, vol. 3729, pp. 624–637. Springer, Heidelberg (2005)
20. Levenshtein, V.: Binary codes capable of correcting deletions, insertions, and reversals. Doklady Akademii Nauk. SSSR 163, 845–848 (1965)
21. Munkres, J.: Algorithms for the Assignment and Transportation Problems. Journal of the Society for Industrial and Applied Mathematics 5, 32–38 (1957)

Part II
Automated Negotiating Agents Competition

The Second Automated Negotiating Agents Competition (ANAC2011)

Katsuhide Fujita, Takayuki Ito, Tim Baarslag, Koen Hindriks, Catholijn Jonker, Sarit Kraus, and Raz Lin

Abstract. In May 2011, we organized the Second International Automated Negotiating Agents Competition (ANAC2011) in conjunction with AAMAS 2011. ANAC is an international competition that challenges researchers to develop a successful automated negotiator for scenarios where there is incomplete information about the opponent. One of the goals of this competition is to help steer the research in the area of bilateral multi-issue negotiations, and to encourage the design of generic negotiating agents that are able to operate in a variety of scenarios. Eighteen teams from seven different institutes competed in ANAC2011. This chapter describes the participating agents and the setup of the tournament, including the different negotiation scenarios that were used in the competition. We report on the results of the qualifying and final round of the tournament.

1 Introduction

Negotiation is an important process to form alliances and to reach trade agreements. Research in the field of negotiation originates from various disciplines including

Katsuhide Fujita
School of Engineering, The University of Tokyo
e-mail: fujita@ipr-ctr.t.u-tokyo.ac.jp

Takayuki Ito
Techno-Business Administration (MTBA), Nagoya Institute of Technology
e-mail: ito.takayuki@nitech.ac.jp

Tim Baarslag, Koen Hindriks, Catholijn Jonker
Man Machine Interaction Group, Delft University of Technology
e-mail: {T.Baarslag,K.V.Hindriks,C.M.Jonker}@tudelft.nl

Sarit Kraus · Raz Lin
Computer Science Department, Bar-Ilan University
e-mail: {linraz,sarit}@cs.biu.ac.il

Sarit Kraus
Institute for Advanced Computer Studies, University of Maryland

T. Ito et al. (Eds.): Complex Automated Negotiations, SCI 435, pp. 183–197.
springerlink.com

economics, social science, game theory and artificial intelligence (e.g., [2, 10, 14]). Automated agents can be used side by side the human negotiator embarking on an important negotiation task. They can alleviate some of the efforts required of people during negotiations and also assist people that are less qualified in the negotiation process. There may even be situations in which automated negotiators can replace the human negotiators. Another possibility is for people to use these agents as a training tool, prior to actually performing the task. Thus, success in developing an automated agent with negotiation capabilities has great advantages and implications.

In order to help focus research on proficiently negotiating automated agents, we have organized the first automated negotiating agents competition (ANAC). The principal goals of the ANAC competition are as follows:

- Encouraging the design of agents that can proficiently negotiate in a variety of circumstances,
- Objectively evaluating different bargaining strategies,
- Exploring different learning and adaptation strategies and opponent models, and
- Collecting state-of-the-art negotiating agents, negotiation domains, and preference profiles, and making them available and accessible for the negotiation research community.

A number of successful negotiation strategies already exist in literature [5, 6, 8, 9]. However, the results of the different implementations are difficult to compare, as various setups are used for experiments in ad hoc negotiation environments [12]. An additional goal of ANAC is to build a community in which work on negotiating agents can be compared by standardized negotiation benchmarks to evaluate the performance of both new and existing agents.

In designing proficient negotiating agents, standard game-theoretic approaches cannot be directly applied. Game theory models assume complete information settings and perfect rationality [15]. However, human behavior is diverse and cannot be captured by a monolithic model. Humans tend to make mistakes, and they are affected by cognitive, social and cultural factors [3, 4, 13]. A means of overcoming these limitations is to use heuristic approaches to design negotiating agents. When negotiating agents are designed using a heuristic method, we need an extensive evaluation, typically through simulations and empirical analysis.

We have recently introduced an environment that allowed us to evaluate agents in a negotiation competition such as ANAC: GENIUS [12], a **G**eneral **E**nvironment for **N**egotiation with **I**ntelligent multi-purpose **U**sage **S**imulation. GENIUS helps facilitating the *design* and *evaluation* of automated negotiators' strategies. It allows easy development and integration of existing negotiating agents, and can be used to simulate individual negotiation sessions, as well as tournaments between negotiating agents in various negotiation scenarios. The design of general automated agents that can negotiate proficiently is a challenging task, as the designer must consider different possible environments and constraints. GENIUS can assist in this task, by allowing the specification of different negotiation domains and preference profiles by means of a graphical user interface. It can be used to train human negotiators by means of negotiations against automated agents or other people. Furthermore, it can be used to teach the design of generic automated negotiating agents.

The Automated Negotiating Agents Competition (ANAC) 2010 was held on May 12, 2010, with the finals being run during the AAMAS 2010 conference. Seven teams have participated in the first competition and *AgentK* generated by the Nagoya Institute of Technology team won the ANAC2010[1]. The tournament was ran on three different domains in ANAC2010 namely: Zimbabwe-England domain, Itex vs Cypress domain and Travel domain. The main differences between ANAC2010 in last year and ANAC2011 in this year are two points: Shared Timeline and Discount Factor. In ANAC 2010, the agents had three minutes each to deliberate. This means agents have to keep track of both their own time and the time the opponent has left. For ANAC2011, we have chosen a simpler protocol where both agents have a shared time window of three minutes. ANAC 2011 has domains that have discount factors. In ANAC 2010, almost every negotiation between the agents took the entire negotiation time of three minutes each to reach an agreement. Adding discount factors should provide more interesting negotiations with faster deals.

The timeline of ANAC2011 is mainly consisted by three parts: Qualifying Round, Updating Period and Final Round. The domains and preference profiles used during the competition are not known in advance and were designed by all participants. First, the qualifying round was played in order to select the best 8 agents from 18 agents. The entire pairwise matches played among 18 agents, and the best 8 agents of those tournaments proceed to the Finals. We set up the updating period for improving the finalists' agents for the final round. The detail results and all domains for the qualifying round are revealed to all finalists, and they tuned up their agents. Time period of updating period is for two weeks. Finally, the final round was played among 8 agents. The domains and preference profiles in the final were 8 domains submitted by all finalists for the final round. The entire pairwise matches played among 8 agents, and the ranking of ANAC2011 is decided.

The remainder of this paper is organized as follows. Section 2 provides an overview over the design choices for ANAC, including the model of negotiation, tournament platform and evaluation criteria. In Section 3, we present the setup of ANAC2011 followed by Section 4 that layouts the results of competition. Finally, Section 5 outlines our conclusions and our plans for future competitions.

2 Set Up of ANAC

2.1 Negotiation Model

Given the goals outlined in the introduction, in this section we introduce the set-up and negotiation protocol used in ANAC. In this competition, we consider *bilateral* negotiations, i.e. negotiation between two parties. The interaction between negotiating parties is regulated by a *negotiation protocol* that defines the rules of how and when proposals can be exchanged. In the competition, we use the alternating-offers protocol for bilateral negotiation as proposed in [16], in which the negotiating parties exchange offers in turns. The alternating-offers protocol conforms with our criterion to have simplicity of rules. Moreover, it is a protocol which is widely studied and used in literature, both in game-theoretic and heuristic settings of negotiation (a non-exhaustive list includes [7, 10, 11, 14, 15]).

Now, the parties negotiate over a set of *issues*, and every issue has an associated range of alternatives or *values*. A negotiation *outcome* consists of a mapping of every issue to a value, and the set, Ω of all possible outcomes is called the negotiation *domain*. The domain is common knowledge to the negotiating parties and stays fixed during a single negotiation session. In ANAC2011, we focused on settings with a finite set of discrete values per issue.

In addition to the domain, both parties also have privately-known preferences described by their *preference profiles* over Ω. These preferences are modelled using a utility function U that maps a possible outcomes $\omega \in \Omega$ to a real-valued number in the range $[0,1]$. In ANAC2011, the utilities are *linearly additive*. That is, the overall utility consists of a weighted sum of the utility for each individual issue. While the domain (i.e. the set of outcomes) is common knowledge, the preference profile of each player is private information. This means that each player has only access to its own utility function, and does not know the preferences of its opponent.[1] Moreover, we use the term *scenario* to refer to the domain and the pair of preference profiles (for each agent) combined.

Finally, we supplement it with a deadline and discount factors. The reasons for doing so are both pragmatic and to make the competition more interesting from a theoretical perspective. Without a deadline, the negotiation might go on forever, especially without any discount factors. Also, with unlimited time an agent may simply try a large number of proposals to learn the opponent's preferences. In addition, as opposed to having a fixed number of rounds, both the discount factor are measured in *real time*. In particular, it introduces yet another factor of uncertainty since it is now unclear how many negotiation rounds there will be, and how much time an opponent requires to compute a counter offer. Also, this computational time will typically change depending on the size of the outcome space. In ANAC2011, the discount factors depend on the scenario, but the deadline is fixed and is set to three minutes, in which both agents shared this fixed time window[2]. The implementation of discount factors in ANAC2011 is as follows. Let d in $[0,1]$ be the discount factor. Let t in $[0,1]$ be the current normalised time, as defined by the timeline. We compute the discounted utility U_D^t as follows:

[1] We note that, in the competition each agent plays *both* preference profiles, and therefore it would be possible in theory to learn the opponent's preferences. However, the rules explicitly disallow learning *between* negotiation sessions, and only *within* a negotiation session. This is done so that agents need to be designed to deal with unknown opponents.

[2] In contrast, in ANAC 2010, the agents had three minutes *each* to deliberate. This means the agents had to keep track of both their own time and the time the opponent had left, otherwise they run the risk of the opponent walking away unexpectedly.

$$U_D^t(s_1, s_2) = U(s_1, s_2) \cdot d^t \tag{1}$$

If $d = 1$, the utility is not affected by time, and such a scenario is considered to be undiscounted, while if d is very small there is high pressure on the agents to reach an agreement. Note that, in the set-up used in ANAC2011 competition, discount factors are part of the preference profiles and are always *symmetric* (i.e. d always has the same value for both agents).

2.2 *Running the Tournament*

As a tournament platform to run and analyse the negotiations, we use the GENIUS environment (General Environment for Negotiation with Intelligent multi-purpose Usage Simulation) [12]. GENIUS is a research tool for automated multi-issue negotiation, that facilitates the design and evaluation of automated negotiators' strategies. It also provides an easily accessible framework to develop negotiating agents via a public API. This setup makes it straightforward to implement an agent and to focus on the development of strategies that work in a general environment.

GENIUS incorporates several mechanisms that aim to support the design of a general automated negotiator. The first mechanism is an analytical toolbox, which provides a variety of tools to analyse the performance of agents, the outcome of the negotiation and its dynamics. The second mechanism is a repository of domains and utility functions. Lastly, it also comprises repositories of automated negotiators. In addition, GENIUS enables the evaluation of different strategies used by automated agents that were designed using the tool. This is an important contribution as it allows researchers to empirically and *objectively* compare their agents with others in different domains and settings.

The timeline of ANAC2011 consists of three phases: the qualifying round, the updating period and the final round. The domains and preference profiles used during the competition are not known in advance and were designed by the participants themselves. An agent's success is measured using the evaluation metric in all negotiations of the tournament for which it is scheduled.

First, a *qualifying round* was played in order to select the best 8 agents from the 18 agents that were submitted by the participating teams. Each participant also submitted a domain and pair of preference profiles for that domain. All 18 such domains were used in the qualifying rounds. For each of these domains, negotiations were carried out between all pairings of the 18 agents.

Since there were 18 agents, which each negotiate against 17 other agents, in 18 different domains, a single tournament in the qualifying round consists of $18 \times 17/2 \times 2 \times 18 = 5508$ negotiation sessions[3]. To reduce the effect of variation in the results, the tournament was repeated 3 times, leading to a total of 16,524 negotiation sessions, each with a time limit of three minutes. In order to complete such an extensive set of tournaments within a limited time frame, we used five high-spec

[3] The combinations of 18 agents are $18 \times 17/2$, however, agents play each domain against each other twice by switching the roles.

computers, made available by Nagoya Institute of Technology. Specifically, each of these machines contained an *Intel Core i7* CPU, at least 4GB of DDR3 memory, and a hard drive with at least 500GB of capacity.

The best 8 agents of those tournaments proceed to the finals round. In the qualifying round, considering all possible pairwise matches among the submitted agents is fairer than randomly dividing agents into groups, because in this way, unfair grouping is avoided (e.g. it avoids the situation that some of the groups could be much more competitive than others). The results from the preliminary tournament matching all submitted agents was used for selecting the best 8 agents taking part in the final round.

Between the 3 rounds, we allowed a 2-week updating period, in which the 8 selected finalists were given the chance to improve their agents for the final round. The detailed results and all domains for the qualifying round were revealed to all finalists, and they could use this additional information to tune their agents. The updating period was set at two weeks.

The final round was played among the 8 agents that achieved the best average scores during qualifying. The domains and preference profiles in the final were 8 domains submitted by all finalists for the final round. The entire pairwise matches played among 8 agents, and the final ranking of ANAC2011 was decided. In the final, a single tournament consists of $8 \times 7/2 \times 2 \times 8 = 448$ negotiation sessions[4]. Again, each tournament was repeated three times.

3 Competition Domains and Agents

3.1 Scenario Descriptions

The ANAC is aimed towards modelling multi-issue negotiations in uncertain, open environments, in which agents do not know what the preference profile of the opponent is.

Table 1 The domains used in ANAC2011 Final Round

Domain Name	Number of issues	Size	Opposition
adg	5	25	Weak
NiceOrDie	1	3	Strong
Energy Domain	8	390,625	Strong
IS_BT_Acquisition	5	384	Medium
Grocery	5	1,600	Weak
Amsterdom_party	6	2,268	Weak
laptopdomain	3	27	Weak
CameraDoamin	6	3,600	Medium

[4] The combinations of 8 agents are $8 \times 7/2$, however, agents play each domain against each other twice by switching the roles.

The various characteristics of a negotiation scenario such as size, number of issues, opposition, discount factor can have a great influence on the negotiation outcome. Therefore, in order to ensure a good spread of negotiation characteristics and fairness, and to reduce any possible bias on the part of the organisers, we gathered the domains and profiles from the participants in the competition. Specifically, in addition to submitting their agents, each participant submitted a scenario, consisting of both a domain and a pair of preference profiles. In the qualifying round, we used all 18 scenarios submitted by the participants. In the final round, eight scenarios submitted by the eight finalists were used. The final scenarios vary in terms of the number of issues, the number of possible proposals, the opposition of the preference profiles and the mean distance of all of the points in the outcome space to the Pareto frontier (see Table 1). The shape of the outcome space of each scenario is presented graphically in Figure 1.

The details of the scenarios are as follows:

Car

The *Car* domain represents a scenario in which a car dealer negotiates with a potential buyer. There are 6 negotiation issues, which represent the features of the car (such as CD player, extra speakers and air conditioning) and each issue takes one of 5 values (good, fairly good, standard, meagre, none), creating 15,625 possible agreements. The domain is almost symmetric and ensures that outcomes with very high utility for both parties can be achieved. Although the best bids of the domain are worth zero for the opponent, this domain is far from a zero sum game. Agents lean to make the agreement which is a 0.85 vs. 0.98 result. An agent can get close to the maximum possible utility (1.00), if it persuades its opponent to accept 0.85. The domain also allows agents to compromise to a fair division point (0.93, 0.93).

Nice Or Die

This domain is very different to the others, due to its very small size and competitiveness. In this domain, agents have to select between 3 possible agreement points: a fair division point, which is less efficient (in the sense that the sum of the agent's utilities is smaller) or one of two selfish points. The domain is symmetric and, naturally, there are only three possible outcomes. The fair division point has utility of (0.29, 0.29), while the other two selfish points have utilities of (1.00, 0.16) and (0.16, 1.00). In the selfish point, one agent can get a high utility despite the utility of the opponent is very low. If agents try to get high utilities, it is hard for them to reach agreements. However, if agents would like to make an agreement in this scenario, the social welfare is small (as, in the ANAC set-up, the agents cannot learn from previous interactions with an opponent).

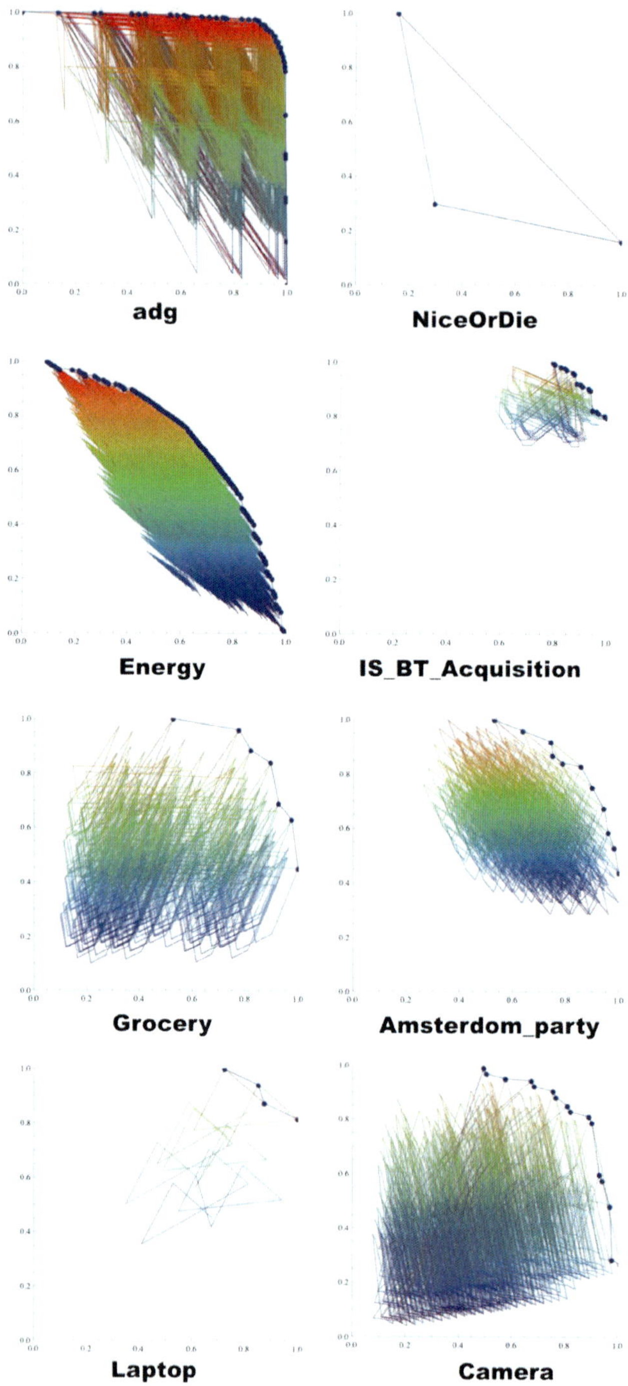

Fig. 1 Acceptance probability space

Energy

This domain considers the problem faced by many electricity companies to reduce electricity consumption during peak times, which requires costly resources to be available and puts a high pressure on local electricity grids. This domain models this application scenario as follows. One agent represents the electricity distribution company whilst the other represents a large consumer. The issues they are negotiating over represent how much the consumer is willing to reduce its consumption over a number of time slots for a day-ahead market (the 24 hours in a day are discretised into 3 hourly time slots). For each issue, there is a demand reduction level possible from zero up to a maximum possible (specifically, 100 kW).

In this domain, the distributor obtains utility by encouraging consumers to reduce their consumptions. Participants set their energy consumption (in kWh) for each of 8 time slots. In each slot, they can reduce their consumption by 0, 25, 50, 75 or 100 kWh. This domain is the largest in the ANAC11 competition (390,625 possible agreements) and has a highly competitive outcome space, therefore, reaching mutually beneficial agreements requires extensive exploration of the outcome space by the negotiating agents.

Company Acquisition

This domain represents a negotiation between two companies, in which the management of Intelligent Solutions Inc. (IS) wants to acquire the BI-Tech company. The negotiation includes five issues: the price that IS pays for BI Tech, the transfer of intellectual property, the stocks given to the BI-Tech founders, the terms of the employees' contracts and the legal liability of Intelligent Solutions Inc.

Each company wants to be the owner of the intellectual property. For IS, this issue is much more important. IS and BI-Tech have common interest that the BI-Tech co-founders would get jobs in IS. IS prefers to give BI-Tech only 2% of the stocks, while the BI-Tech co-founders want 5%. IS prefer private contracts, while firing workers is less desirable by them. BI-Tech prefers a 15% salary raise. For both sides this is not the most important issue in the negotiation. Each side prefers the least legal liability as possible.

The utility range is narrow and has high utility values such that all outcomes give both participants a utility of at least 0.5. The domain is relatively small, with 384 possible outcomes.

Grocery

The *Grocery* domain models a discussion in a local supermarket. The negotiation is between two people living together who have different tastes. The discussion is about five types of product: bread, fruit, snacks, spreads, and vegetables. Each category consists of four or five products, resulting in a medium sized domain with 1,600 possible outcomes. For their daily routine it is essential that a product of each type is present in their final selection, however only one product can be selected for each type. Besides their difference in taste, they also differ in what category of product they find more important.

The profiles for agents Mary and Sam are modelled in such a way that a good outcome is achievable for both. Sam has a slight advantage, since he is easier to satisfy than Mary, and therefore is likely to have better outcomes. This scenario allows outcomes that are mutually beneficial, but the outcome space is scattered so agents must explore it considerably to find the jointly profitable ones.

Amsterdam Trip

The domain concerns the planning of a touristic trip to Amsterdam and includes issues representing the day and time of travel, the duration of the trip, the type of venues to be visited, the means of transportation and the souvenirs to buy. This domain is moderately large as the utility space has 3,024 possible bid configurations. The preference profiles specify a generous win-win scenario, since it would be unrealistic for two friends to make a trip to Amsterdam and to have it be a zero-sum game.

The size of the domain enables the agent to communicate their preferences (by means of generating bids), without having to concede far. Also the magnitude of the domain puts agents which use a random method of generating bids at a disadvantage, since the odds of randomly selecting a Pareto optimal bid in a large domain are small. So this domain will give an advantage to agents that make some attempt to learn the opponents preference profile, and those capable of rapidly choosing offers.

Laptop

This domain is a variation of the *Laptop* domain from the qualification rounds. Two parties, a seller and a buyer, are negotiating the specifications of a laptop. The domain has three issues: The laptop brand, the size of the hard disk, and the size of the external monitor. Each issue has only three options, making it a very small domain of 27 possible outcomes.

For example, in a negotiation about buying a laptop the buyer may prefer to have a middle-sized screen but the seller may prefer to sell laptops with small screens because s/he has more of those in stock. They could, however, agree on the brand of laptop that they want to buy/sell. An outcome of a negotiation reconciles such differences and results in a purchase.

Camera

This domain is another retail domain, which represents the negotiation between a buyer and a seller of a camera. It has six issues: makers, body, lens, tripods, bags, and accessories. The size of this domain is 3,600 outcomes. The seller gives priority to the maker, and the buyer gives priority to the lens. The competitiveness of this negotiation domain is medium.

The range of the contract space is wide, which means the agents need to explore it to find the jointly profitable outcomes. While jointly profitable outcomes are possible (since the Pareto frontier is concave), no party has an undue advantage in this (since the Nash point is at an impartial position).

3.2 Agent Descriptions

ANAC2011 had 18 agents, registered from 7 universities: Bar Ilan University, Israel (×5); University of Southampton, United Kingdom (×2); Ben-Gurion University, Israel (×4); Delft University of Technology, The Netherlands (×4); Politehnica University of Bucharest, Romania; University of Alcalá, Spain and Nagoya Institute of Technology, Japan (one each).

Table 2 Team Members in the Final Round

No.	Team Members	Affliction	Agent Name	Domain Name
1	Asaf Frieder Dror Sholomon Gal Miller	Bar Ilan University	ValueModelAgent	adg
5	Mai Ben Adar Nadav Sofy Avshalom Elimelech	Bar Ilan University	Gahboninho	NiceOrDie
6	Colin R. Williams Valentin Robu	University of Southampton	IAMhaggler2011	Energy Domain
8	Radmila Fishel Maya Bercovitch, Ayelet Urieli Betty Sayag	Ben-Gurion University	BRAMAgent	IS_BT_Acquisition
12	Alex Dirkzwager Mark Hendrikx Julian de Ruiter	TU Delft	TheNegotiator	Grocery
13	Thijs van Krimpen Daphne Looije Siamak Hajizadeh	TU Delft	HardHeaded	Amsterdom_party
16	Tim Baarslag Koen Hindriks Catholijn Jonker	TU Delft	Nice Tit for Tat agent	laptopdomain
18	Shogo Kawaguchi Katsuhide Fujita Takayuki Ito	Nagoya Institute of Technology	AgentK2	CameraDoamin

The final round in ANAC2011 had eight teams from four different universities, as listed in Table 2. They are the winners of the qualifying round. In the rest of the chapter in this book, we provide sections of the individual strategies of the ANAC2011 finalists based on descriptions of the strategies provided by the teams.

4 Competition Results

The ANAC11 competition consisted of two rounds: a qualifying round and a final round. We describe the results of these rounds in turn.

Table 3 Average scores of every strategy in the qualifying round

Rank	Agent Strategy	Mean Utility
1	Gahboninho	0.756
2	HardHeaded	0.708
3	ValueModelAgent	0.706
4	AgentK2	0.702
5	IAMhaggler2011	0.701
6	BRAMAgent	0.690
7	Nice Tit-For-Tat Agent	0.686
8	TheNegotiator	0.685
9	GYRL	0.678
10	WinnerAgent	0.671
11	Chameleon	0.664
12	SimpleAgentNew	0.648
13	LYYAgent	0.640
14	MrFriendly	0.631
15	AgentSmith	0.625
16	IAMcrazyHaggler	0.623
17	DNAgent	0.601
18	ShAgent	0.571

4.1 Qualifying Round

The qualifying round consisted of 18 agents that were submitted to the competition. For each pair of agents, under each preference profile, we ran a total of 3 negotiations. By averaging over all the scores achieved by each agent (against all opponents and using all preference profiles), eight finalists were selected based on their average scores. Formally, the average score $U_\Omega(p)$ of agent p in scenario Ω is given by:

$$U_\Omega(p) = \frac{\sum_{p' \in P, p \neq p'} U_\Omega(p, p') + U_\Omega(p', p)}{2 \cdot (|P| - 1)} \tag{2}$$

where P is the set of players and $U_\Omega(p, p')$ is the utility achieved by player p against player p' when player p is under the side A of Ω and player p' is under the side B of Ω. It is notable that *Gahboninho* was the clear winner of the qualifying round (see Table 3). However, the differences in utilities between many of the middle ranked strategies are rather small, so several of the agents which qualified for the final only did so by a small margin.

4.2 Final Round

For the final round, we matched each pair of finalist agents, under each preference profile, a total of 3 times. Participants were given the opportunity to submit revised agents for the final based on the results of the qualifying round. Table 4 summarises the means, standard deviations, and 95% confidence interval bounds for the results

Table 4 Tournament results in the final round

Agent	Score	SD	low CI	high CI
HardHeaded	0.748697121	0.00956806	0.74512475	0.752269492
Gahboninho	0.739685706	0.005244734	0.737727511	0.741643902
IAMhaggler2011	0.686419417	0.004663924	0.684678075	0.688160759
AgentK2	0.680806502	0.004701551	0.679051111	0.682561893
TheNegotiator	0.680365635	0.004348087	0.678742215	0.681989055
BRAMAgent	0.679967769	0.005026365	0.678091104	0.681844434
Nice Tit for Tat Agent	0.678179856	0.007599692	0.675342403	0.68101731
ValueModelAgent	0.616818076	0.006890059	0.614245575	0.619390578

Table 5 Detailed scores of every agent in the final round

Agent	adg	NorD	Energy	IS_BT	Grocery	AMS	laptop	Camera
HardHeaded	0.954	0.500	0.549	0.744	0.724	0.867	0.664	0.805
Gahboninho	0.942	0.511	0.676	0.751	0.673	0.914	0.745	0.671
IAMhaggler2011	0.872	0.300	0.558	0.824	0.741	0.787	0.767	0.727
AgentK2	0.922	0.375	0.483	0.797	0.727	0.762	0.673	0.734
TheNegotiator	0.931	0.317	0.533	0.749	0.732	0.797	0.641	0.745
BRAMAgent	0.822	0.500	0.452	0.737	0.724	0.795	0.608	0.741
Nice Tit for Tat Agent	0.784	0.445	0.518	0.754	0.745	0.750	0.630	0.756
ValueModelAgent	0.941	0.193	0.326	0.748	0.758	0.852	0.607	0.777

of each agent. In common with the approach used in the qualifying round, all agents use both of the profiles that are linked to a scenario. The average score achieved by each agent in each scenario is given in Table 5. In the finals, *HardHeaded* proved to be the clear winner, with a score of 0.749.

In more detail, we can consider the performance of the agents in each scenario. Table 5 gives the average score of each agent in each scenario. It shows that *HardHeaded* and *Gahboninho* won by a big margin in most of scenarios. Usually, some agents lost the utility in scenarios with a large size and high opposition, however, *HardHeaded* and *Gahboninho* could get the higher utility in such "tough" scenarios. In addition, *IAMhaggler2011* won the *Company Acquisition* and *Laptop* scenarios with low discount factor, therefore, *IAMhaggler2011* has a high capacity to the discount factors. The differences among *BRAMAgent*, *AgentK2*, *TheNegotiator* are very small.

5 Conclusion

This paper describes the second automated negotiating agents competition. Based on the process, the submissions and the closing session of the competition we believe that our aim has been accomplished. Recall that we set out for this competition in order to steer the research in the area bilateral multi-issue closed negotiation. The competition has achieved just that. Eighteen teams have participated in the competition and we hope that many more will participate in the following competitions.

One of the successes of ANAC lies in the development of state-of-the-art negotiation strategies that co–evolve every year. This incarnation of ANAC already yielded seven new strategies and we hope that next year will bring even more sophisticated negotiation strategies. ANAC also has an impact on the development of GENIUS. We have released a new, public build of GENIUS[5] containing all relevant aspects of ANAC. In particular, this includes all domains, preference profiles and agents that were used in the competition. This will make the complete setup of ANAC available to the negotiation research community.

Not only have we learnt from the strategy concepts introduced in ANAC, we have also gained understanding in the correct setup of a negotiation competition. The joint discussion with the teams gives great insights into the organizing side of the competition.

To summarize, the agents developed for ANAC are the first step towards creating autonomous bargaining agents for real negotiation problems. We plan to organize the second ANAC in conjunction with the next AAMAS conference in 2012.

Acknowledgements. The authors would like to thank the team of masters students at Nagoya Institute of Technology, Japan for their valuable help in the organisation of the ANAC2011 competition. Moreover, the authors acknowledge the use of the IRIDIS High Performance Computing Facility, and associated support services at the University of Southampton, in the completion of this work. Furthermore, this research is supported by the Dutch Technology Foundation STW, applied science division of NWO and the Technology Program of the Ministry of Economic Affairs. It is part of the Pocket Negotiator project with grant number VICI-project 08075. In addition, this research is based upon work supported in part under NSF grant 0705587, by the U. S. Army Research Laboratory and the U. S. Army Research Office under grant number W911NF-08-1-0144 and by ERC grant #267523.

References

1. Ito, T., Zhang, M., Robu, V., Fatima, S., Matsuo, T. (eds.): New Trends in Agent-Based Complex Automated Negotiations. SCI, vol. 383. Springer, Heidelberg (2012)
2. Aumann, R.J., Hart, S. (eds.): Handbook of Game Theory with Economic Applications, vol. 1. Elsevier (March 1992)
3. Bazerman, M.H., Neale, M.A.: Negotiator rationality and negotiator cognition: The interactive roles of prescriptive and descriptive research. In: Young, H.P. (ed.) Negotiation Analysis, pp. 109–130. The University of Michigan Press (1992)
4. Erev, I., Roth, A.: Predicting how people play games: Reinforcement learning in experimental games with unique, mixed strategy equilibrium. American Economic Review 88(4), 848–881 (1998)
5. Faratin, P., Sierra, C., Jennings, N.R.: Negotiation decision functions for autonomous agents. Int. Journal of Robotics and Autonomous Systems 24(3-4), 159–182 (1998)
6. Faratin, P., Sierra, C., Jennings, N.R.: Using similarity criteria to make negotiation tradeoffs. Journal of Artificial Intelligence 142(2), 205–237 (2003)
7. Fatima, S.S., Wooldridge, M., Jennings, N.R.: Multi-issue negotiation under time constraints. In: AAMAS 2002: Proceedings of the First International Joint Conference on Autonomous Agents and Multiagent Systems, pp. 143–150. ACM, New York (2002)
8. Ito, T., Hattori, H., Klein, M.: Multi-issue negotiation protocol for agents: Exploring nonlinear utility spaces (2007)

[5] http://mmi.tudelft.nl/negotiation/index.php/Genius

9. Jonker, C.M., Robu, V., Treur, J.: An agent architecture for multi-attribute negotiation using incomplete preference information. Journal of Autonomous Agents and Multi-Agent Systems 15(2), 221–252 (2007)
10. Kraus, S.: Strategic Negotiation in Multiagent Environments. MIT Press (October 2001)
11. Kraus, S., Wilkenfeld, J., Zlotkin, G.: Multiagent negotiation under time constraints. Artificial Intelligence 75(2), 297–345 (1995)
12. Lin, R., Kraus, S., Tykhonov, D., Hindriks, K., Jonker, C.M.: Supporting the Design of General Automated Negotiators. In: Ito, T., Zhang, M., Robu, V., Fatima, S., Matsuo, T., Yamaki, H. (eds.) Innovations in Agent-Based Complex Automated Negotiations. SCI, vol. 319, pp. 69–87. Springer, Heidelberg (2010)
13. McKelvey, R.D., Palfrey, T.R.: An experimental study of the centipede game. Econometrica 60(4), 803–836 (1992)
14. Osborne, M.J., Rubinstein, A.: Bargaining and Markets (Economic Theory, Econometrics, and Mathematical Economics). Academic Press (April 1990)
15. Osborne, M.J., Rubinstein, A.: A Course in Game Theory. MIT Press (1994)
16. Rubinstein, A.: Perfect equilibrium in a bargaining model. Econometrica 50(1), 97–109 (1982)

Value Model Agent: A Novel Preference Profiler for Negotiation with Agents

Asaf Frieder and Gal Miller

Abstract. As multi-agent systems become more complicated, situations arise were agents have different goals. Accomplishing their goals may require trading resources or compromising on the state of the system. In these situations cooperation can improve the utilities of all parties, but some agreements are preferred to others. Humans have reached these agreements for thousands of years through an art known as negotiation. In this paper we describe an agent that was developed for the ANAC2011 bilateral negotiation competition. Our main contribution is a novel approach to modeling the preference profile of the other agent. This preference profile is then used to improve exploration of the bid space and approximate the opponent's concessions.

1 Introduction

For thousands of years humans have resolved conflicts using negotiation. From children deciding what to play to business deals and peace agreements, negotiation is an integral part of human interaction. In some multi-agent systems much like in a community, agents have different goals or preferences. In these cases the agents may conflict over their need of resources or their desired world state. The agents may interfere with each others' plans by withholding necessary resources or changing the world state. It can therefore be beneficial for agents to agree on a common coordinated plan that is more efficient for both agents even if not optimal for either. However, in most negotiation domains there are multiple possible plans (bids) that can be chosen and the agents have different preferences. Thus, an agent needs negotiation skills to ensure its own preferences will be prioritized in the final agreement.

The *ValueModelAgent* was designed to negotiate with other agents as part of the ANAC2011 negotiation competition. Our primary effort was focused on developing a heuristic for reconstructing the profile of the other agent's preferences described in Section 2. An accurate model of the opponent's preferences has many possible

T. Ito et al. (Eds.): Complex Automated Negotiations, SCI 435, pp. 199–203.
springerlink.com

applications in negotiation. The model can be used to increase the likelihood of the opponent accepting bids, by choosing the bids our opponent prefers. The preference profile can also be used to calculate reservation values based on measurements of fairness such as the nash and kalai points. The opponent's concession rate can also be calculated using a preference profile. Using the concession rate, the agent can model the opponent's behavior and estimate the future agreement under different bidding strategies.

We farther describe our bidding strategy that uses the profile to select bids as well as approximate our opponent's concession in Section 3. Our bidding strategy is rule based and takes both our opponent's concessions and the timeline into account.

2 Modeling Preferences

The ValueModel of our agent attempts to extrapolate the opponent's preference profile from the order of our opponent's bids. However, our heuristic requires utility approximations for our opponent's bids. We therefore must generate initial approximations for these bids based on their order. The *a priori* approximations for the opponent's bids are generated based on the assumptions that the opponent makes bids in the order of their preference profile and that the preference profiles of both parties are roughly symmetric. Under these assumptions, the *a priori* approximation for the i-th bid proposed by our opponent is approximate to the utility of the i-th highest bid in our preference profile. While the inaccuracies in these assumptions may cause noise in the *a priori* approximations, they serve only as a starting point for the heuristic we will now describe.

In the genius platform a bid B is a set of values chosen for each of N issues ($B = v_1..v_N$). Recall that a negotiation domain has several issues, each with a different weight w_I for our opponent. Each issue has several values, with different scores. The ratio between a value's score and the maximal score in the issue, is the portion of the issue's weight that our opponent will gain if that value were to be chosen. We define value utility loss (VUL) as the maximal difference between our opponent's utility for two bids if one contains the value and the other can only differ in the value for that issue ($I(v)$) (see Eq. 1). We farther define the bid utility loss (BUL) as the sum of the utility lost by all values of the bid (or concession).

$$VUL(v) = w_{I(v)} \cdot \left(1 - \frac{eval(v)}{\max\limits_{v' \in I(v)} eval(v')} \right) \tag{1}$$

In the value model, each value of each issue has three properties, the VUL approximation, the reliability (rel) the agent associates to that approximation and a weighted deviation (sd) measurement. When we receive the opponent's first bid, we assume the values in that bid are optimal ($\forall i, VUL(v_i) = 0$), with very high reliability (0.9). We also initialize all other values with high utility loss ($VUL = 1/N$), a very low reliability (0.02) and high deviation ($1/N$). We will now describe how we update these properties when we receive new bids.

Let $BUL(B)$ be the *a priori* approximation for the utility lost by our opponent's new bid. For each value v_i in bid B our model already has an approximation $VUL(v_i)$, and the sum of these approximation is an alternative approximation for $BUL(B)$. Our heuristic needs to be updated based on the difference between the approximations. Unfortunately the model's approximation for *BUL* is derived from several *VUL* variables and we need to decide which values need to change. We propose that the values with the lowest reliability and highest deviation are most likely responsible for the difference. Equation 2 demonstrates the calculation of the number of units the set of values should move. We then generate new approximations for each value, using Equation 3 such that the sum of these approximations is the a priori *BUL* approximation. In addition we generate a reliability measurement for these approximations using a norm formula on the (inverse) original reliabilities of the bid's values as demonstrated in Equation 4.

$$units = \frac{BUL(B) - \sum_{i=0} VUL(v_i)}{\sum_{i=0}^{n} \frac{1}{\sqrt{rel(v_i)}} \cdot sd(v_i)} \tag{2}$$

$$VUL^B(v_i) = VUL(v_i) + units \cdot sd(v_i) \cdot \frac{1}{\sqrt{rel(v_i)}} \tag{3}$$

$$rel^B(v_i) = \sqrt{\frac{1}{n} \cdot \sum_{i=0}^{n} \frac{1}{rel(v_i)^2}} \tag{4}$$

We are now left with two sets of *VUL* approximations and reliabilities for each value: the model's prior approximations and the ones generated for the new bid. Our objective is to merge these approximations, taking their reliabilities into account such that reliable values will be updated based on temporal difference learning using a learning rate $\alpha = 0.1$ and unreliable values will be updated faster. Therefore, the new approximation takes up to α from the weight (*rel*) of the original approximation and also takes a portion of the weight not claimed by the original approximation ($1 - rel$). In both cases the portion of the weight received by the new approximation depends on its reliability (rel^B). The resulting weight formulas are displayed in Equation 5 and 6, respectably.

$$p = rel(v_i)(1 - \alpha \cdot rel^B(v_i)) \tag{5}$$

$$p^B = rel^B(v_i)(1 - rel(v_i) + \alpha rel(v_i)) \tag{6}$$

We now update the three properties of the value. We first set the value's *VUL* approximation to be the weighted average of the two approximations (see Eq 7). Then we estimate a weighted standard deviation based on the distance of VUL^B and the original distribution (based on *VUL* and *sd*) from the new approximation VUL'.

Finally, we update the reliability of the approximation based on the weights and how accurately the approximation represents VUL and VUL^B (See Eq. 8).

$$VUL'(v_i) = \frac{p \cdot VUL(v_i) + p^B \cdot VUL^B(v_i)}{p + p^B} \tag{7}$$

$$rel' = p\left(1 - \frac{|VUL(v_i) - VUL'(v_i)|}{2sd'(v_i)}\right) + p^B\left(1 - \frac{|VUL^B(v_i) - VUL'(v_i)|}{2sd'(v_i)}\right) \tag{8}$$

3 Bidding Strategy

At any point in the negotiation the agent has a minimal utility threshold and permits itself to propose any bid above that threshold. As long as there are unproposed bids above the threshold in 80% of the bids we propose the unproposed bid with the highest utility for the opponent according to the value model. In 20% of the bids we pick bids with values that were not contained in recent bids. In cases all bids above the threshold were already proposed, we randomly pick a bid from the bids with the highest 25% utilities for the opponent.

Our threshold strategy depends on several factors[1] including the elapsed time. In the beginning of the negotiation we set our threshold to be 0.98. Until 80% of the time elapses we employ a very strict concession strategy. We only lower the threshold if our opponent conceded 50% more than us based on the average (BUL) of the last 5 new bids our opponent proposed. The second requirement is that our opponent has stopped conceding recently. If both requirements are met we gradually lower the threshold by up to 0.02 over a 5% interval of the timeline.

After 80% of the time has elapsed we measure the utility difference between our current threshold and the opponent's highest bid. For the next 10% of the timeline we gradually lower our threshold by up to 50% of this difference but no lower 0.7. than if our . We also restrict our concession to be no more than 0.05 beyond our opponent's concession. We also require that our opponent has stopped making significant concessions.

After 90% of the time elapsed, we try to use scare tactics to convince our opponent to accept our bids by masquerading as ultimatums. We intentionally sleep three times, for half the remaining time, followed by the approximated best bid for our opponent above our threshold. After each of these bids we lower our threshold and resume exploring bids for a short amount of time. If the opponent did not accept our first proposal, we lower our threshold as low as 0.65 and after the second we lower it to 0.6 [2]. If our opponent rejected the third ultimatum, and our opponent's best bid is above 0.55 we propose it. If our opponent's best bid is bellow 0.55, we resume exploring bids above the threshold.

[1] Due to space restrictions we do not mention some of these factors, most notably the discount factor.

[2] In both cases we lower our threshold by only a portion of the utility difference.

4 Conclusions

Our approach is not without faults, and our agent was in eighth place (eighth) in the finals (third in the qualifying rounds). While reconstructing a preference profile may have interesting applications we did not leverage our model much in our bidding strategy. Identifying the best bids for the opponent may increase the chance of the ultimatums being accepted, but its primary use in the exploration is futile since in most cases there is enough time to propose all bids. Using the approach to approximate our opponent's concession has more potential, but was not fully utilized. The effort that was invested in profiling our opponent's preferences would have been much better spent in profiling our opponent's strategy and adjusting our strategy to our opponent's type. For this end we could have measured his concessions using the existing bayesian model, or even the symmetry assumption.

Finally, making serious concessions after only 80% of the time passed is a mistake. Agents can send many proposals in a very short amount of time, and agents that employ hard bargaining strategies will likely wait longer and take advantage of our "early" concessions. An agent that waits longer, can then utilize its preference model to rapidly send the best bid for the opponent under different thresholds.

Gahboninho: Strategy for Balancing Pressure and Compromise in Automated Negotiation

Mai Ben Adar, Nadav Sofy, and Avshalom Elimelech

Abstract. "Gahboninho" agent was first introduced in the ANAC'11 competition. The underlying strategy it implemented follow the assumption it will be matched with other automated agents which vary in their compromising level. Thus, it tries to tackle strong opposition by putting pressure on the opponent. The stubbornness of the agent is also balanced based on the behavior of its opponent in order to achieve higher utilities.

1 Introduction

Our agent (*Gahboninho*) was submitted to the Automated Negotiation Agents Competition (ANAC'11) held in the international Joint Conference on Autonomous Agents and Multi-Agent Systems (AAMAS2011). The agent implemented an original strategy and has reached the second place in the tournament. The strategy itself was motivated by the results of a preliminary experiment with automated agents we took part of, in which the most successful strategy was to decide whether the opponent is susceptible to pressure or is entirely stubborn, then to act rationally. The paper describes our attempt to translate that approach into the competition's settings.

2 Gahboninho's Adaptation Strategy

The agent begins the negotiation without compromising at all in an attempt to quickly pressure the opponent as much as possible. In fact, if it would not pose a threat to the agent's own utility, it may be worthwhile for the agent to propose

---

Mai Ben Adar · Nadav Sofy · Avshalom Elimelech
Bar-Ilan University, Israel

T. Ito et al. (Eds.): Complex Automated Negotiations, SCI 435, pp. 205–208.
springerlink.com

the worst possible outcome for the opponent instead of the best outcome for itself. While such actions may prevent the opponent from accurately constructing the agent's preference profile, the expected rounds count often permits[1] expending many rounds before sending realistic suggestions without the risk of missing Pareto-efficient outcomes.

As the negotiation continues, the agent examines the opponent's behavior. If the opponent expresses an effort[2] to find a realistic and compromising outcome (instead of insisting on certain minimal utility threshold) then unless the time constraints or the discount factor pressure greatly, there is no reason to compromise significantly as it is not likely to drive the opponent to suggest better outcomes. At the same time, the agent may also accurately model the opponent's preferences. Since the opponent's attempt of modeling the agent's preferences is limited to little information, the outcomes suggested by the opponent rely almost entirely on its own preferences thus expected to be genuine. On the other hand, if the opponent will not compromise at all, the agent avoids prolonging the deadlock and gives up utility in accordance to the pressure. As the pressure rises the agent may give up his utility faster, but would never go below the utility of the best outcome suggested by the opponent and neither below a certain pre-determined constant which depends on the amount of competitors in the tournament. Once the agent has enough information about the opponent's preferences or is either pressured enough, it will try filtering the domain of most of the inefficient outcomes in order to insure that the critical, last rounds are relevant and effective.

3 Facing Varied Opponents

Facing a compromising opponent, which avoid expending rounds and rarely insists on a certain utility threshold, is relatively simple as this behavior may be easily identified and exploited by the agent's strategy. Predict the outcome when facing uncompromising opponents is harder, since the reason behind that behavior may be one of the following:

i. The opponent is greedy, gives up his own utility using a predetermined curve and might suggest somewhat compromising outcomes only in the last, pressured rounds. While in this case the agent's strategy copes and practically gives up, which highly benefits the opponent, such an opponent is also much more likely to reach a break-off when facing irrational opponents.

[1] For example, In the competition's qualifying round the average rounds count per negotiation is about 7400, while the domains' possible outcomes count, excluding "energy distributer" domain, averages on 1975 only (the energy domain's was unproportionate and including it turns the average to 24836)

[2] Such estimation may be done, for example, using a heuristic over both the agent's personal utility from the incoming proposals, and the variance of the amount of different issue values in each of the incoming and outgoing outcome proposals (as implemented in our agent)

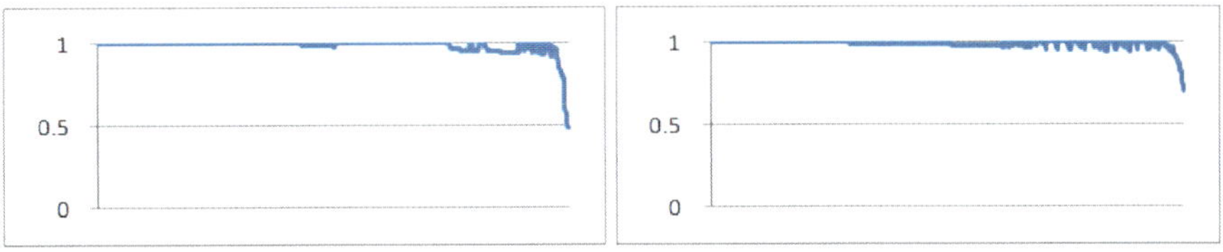

(a) (b)

Fig. 1 Illustration of the utility of our agent's proposals as a function of point in time when negotiating with exemplary opponent which is (a) using a predetermined utility curve (b) rewarding cooperation but keeps insisting in other cases

ii. The opponent's strategy is similar to the agent's. In this case, the negotiation will turn into a hawk-dove game where the fine tuning of the agents determines its result.
iii. The opponent rewards cooperation, but insists if the agent does not compromise. In this case, each time the agent gets pressured a new starts cycle where both the agent and the opponent reveal some more of their preferences by suggesting more compromising outcomes, then the agent returns to insisting, and then the opponent returns to insist as well. Such opponents may reach break-off when facing either greedy agents, or other agents similar to themselves.

Considering all cases, the agent's success in a competition is threatened if there are very few greedy opponents (so their benefit outweighs their break-offs' penalty), or if there is another similar agent(s) which is slightly less compromising.

4 NiceOrDie Domain

As part of the competition, we were required to design and provide a negotiation domain in addition to the agent itself. The purpose of our domain (*NiceOrDie)* was separating the highly greedy agents from the rest, since they pose the greatest threat to the agent, as depicted in the previous section. The domain offers only three possible outcomes. Therefore, the ability to reach an agreement in this domain depends only on the agent's final, minimal personal utility threshold when facing a seemingly greedy opponent. This splits the agents into two groups – those that benefit from each other by compromising, who gain the highest possible utility, and those that won't reach an agreement and will fail at this domain entirely.

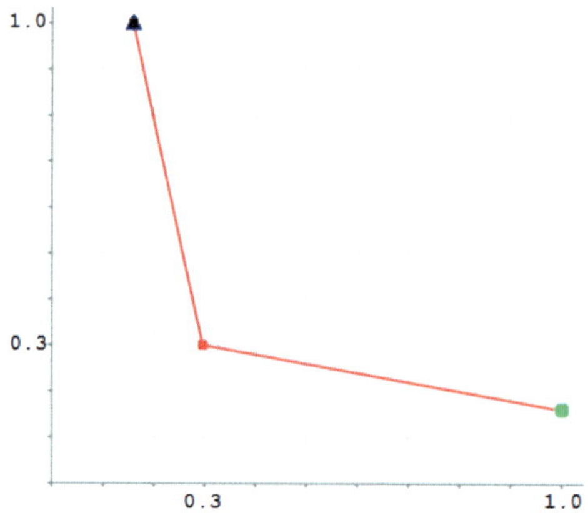

Fig. 2 Illustration of NiceOrDie domain's Pareto-efficient frontier

5 Results and Conclusion

Our agent has reached the first place on the qualification round of the competition and dropped to the second place in the final round. We attribute the agent's success to its ability to be as greedy as a compromising agent can be, thus exploiting compromising opponents as well as most greedy agents, while avoiding break-offs when facing uncompromising opponents. This is evident in the competition's detailed qualification round result, where only two of the eight leading agents[3] reached less break-offs than our agent, and these two agents were highly exploited by our agent during the tournament, as they probably were far less greedy. As described in previous sections, the strategy's success also relies on the participating opponent. Our assumption when designing the agent's strategy was that most competitors would avoid the greedy approach since it is predictable. However, in fear the opponent had chosen that approach exactly, and considering the fact it is simple to identify and counter such an opponent, we assumed that most of the participants will take measures to adapt themselves to such opponents. It is hard to assess how many agents were exploited by our agent's app roach, but in retrospect, it appears that the assumption was fairly accurate. It is worth mentioning the agent implemented a relatively simple opponent model. Additionally, discount factors, which were more significant in the final round than in the qualification round, were handled poorly by the agent's strategy since the pace of the opponent's compromise was not taken in consideration. This may explain the agent's drop in the final round of the competition. Accordingly, extending the agent's strategy and implementation in these areas may increase its success even further.

[3] Gahboninho had 70 break-off, IAMhaggler2011 had 61 and BRAMAgent had only 38 break-offs

IAMhaggler2011: A Gaussian Process Regression Based Negotiation Agent

Colin R. Williams, Valentin Robu, Enrico H. Gerding, and Nicholas R. Jennings

Abstract. We describe the strategy used by our agent, IAMhaggler2011, which finished in third place in the 2011 Automated Negotiating Agent Competition. A key feature of this agent is the way in which it models the likely negotiation behaviour of its opponent. Specifically, it first uses a Gaussian process regression technique to estimate the future concession of its negotiation opponent. Its concession is then set as a best response to this prediction.

1 Introduction

In this work, we give a brief overview of the core parts of the strategy used by our agent, *IAMhaggler2011*, which finished in third place in the second international Automated Negotiating Agent Competition (ANAC2011). The competition setup is described in [1]. The overall framework used by the agent is based on that of our previous (2010) competition entry, *IAMhaggler*[3]. A more detailed description of *IAMhaggler2011* and an analysis of its performance can be found in [4].

The core of the strategy used by *IAMhaggler2011* consists of three parts, which we describe in turn. The first predicts the concession of the opponent (Section 2). The second sets the concession rate such that it optimises the expected utility given that prediction (Section 3). The final part generates a multi-issue offer according to the concession rate (Section 4).

2 Predicting the Opponent's Concession

In order to set its behaviour as a best response to that of the opponent, our agent first needs to predict how the opponent will concede throughout the negotiation, using

Colin R. Williams · Valentin Robu · Enrico H. Gerding · Nicholas R. Jennings
School of Electronics and Computer Science, University of Southampton,
University Road, Southampton, SO17 1BJ
e-mail: {crw104,vr2,eg,nrj}@ecs.soton.ac.uk

T. Ito et al. (Eds.): Complex Automated Negotiations, SCI 435, pp. 209–212.
springerlink.com

only the information which it can observe: the offers made by the opponent, and the utility of these offers according to our agent's utility function. This is done using a Gaussian process regression technique, in order to provide both a prediction of the opponent's future concession and a measure of the level of confidence in that prediction (see [2] for details).

The output of the Gaussian process is a Gaussian distribution, for each time t, denoted by $f(u;\mu_t,\sigma_t)$, where the mean, μ_t, gives an indication of the most likely value for u at time t, whilst the standard deviation, σ_t is an indication of how accurate the prediction of μ_t is likely to be.

As input to the Gaussian process, *IAMhaggler2011* uses the maximum value offered by the opponent in a particular time window of a fixed duration, and the time of that window. This windowed approach reduces the effect of noise on the Gaussian process and reduces the amount of input data. The maximum value in each time window is used, rather than the average, as the maximum represents the best offer that has been observed, and therefore that our agent can expect to reach agreement at.

3 Setting the Concession Rate

We now show how the prediction of the opponent's future concession is used by our strategy to set its concession rate by optimising the expected utility given that prediction. Specifically, we discuss the way in which both the mean, μ, and standard deviation, σ, output by the Gaussian process are used in setting an optimal concession rate. Optimal in this context means that our strategy maximises the agent's expected utility given its prediction of the opponent. The aim is to calculate the best time, t^* and utility value, u^* at which to reach agreement. We consider the best time, t^*, to be the point in future time ($t \in [t_c, t_{max}]$) at which the expected utility, $E_{rec}(t)$ of the opponent's offer is likely to be maximised. The expected utility of reaching an agreement at time t is given by $E_{rec}(t) = \int_0^1 D(R(u)f(u;\mu_t,\sigma_t),t)du$ where $D(\cdot,\cdot)$ is the effect of the discounting factor, $R(\cdot)$ is the risk function, and $f(\cdot)$ is the probability distribution over the values of u, as given by the Gaussian process.[1]

Our strategy considers the risk associated with reaching agreement at a particular utility. In a tournament setting, such as the one used in ANAC2011, the primary aim of the strategy is to 'win' the negotiation by achieving a higher utility than that of its opponent. In such a setting, reaching an agreement with a high utility is not always good enough, since it is possible that the opponent may have achieved an even higher utility, thereby winning the negotiation. Consequently, the strategy may need to take a more aggressive approach than it would if it were simply maximising its own utility.

[1] Due to the constraints on the utility functions used in the competition, we assume that the utility of the opponent's offers must lie in the range $[0,1]$. Therefore we use a truncated normal distribution, with the utility constrained to fit in that range.

In order to deal with this trade-off, we adjust the behaviour of our agent by including the concept of risk attitude in the design of our concession strategy. Formally, the risk function used is $R(u) = u^r$ where r is the risk parameter. If $r = 1$, the strategy is risk-neutral, for $r > 1$, the strategy is risk-seeking and for $r < 1$, it is risk-averse. A risk-seeking strategy would result in more aggressive behaviour, since such an agent will concede more slowly, as it will regard lower utilities to be of even lower value than their true value. This has the effect of making the agent more likely to win in a tournament setting, and is therefore more appropriate in the negotiation competition.

Thus, the effect of the standard deviation, σ_t is as follows. If, at two points in time, t_1 and t_2, the mean values are the same ($\mu_{t_1} = \mu_{t_2}$), but the standard deviation differs such that $\sigma_{t_1} < \sigma_{t_2}$, then a risk-seeking agent will consider the expected utility at time t_2 to be greater than at time t_1. That is, the risk-seeking agent is prepared to wait for the less certain offer at time t_2, as there is a higher chance that the utility may exceed the value of μ_{t_2}, than it would for μ_{t_1}. The converse applies for risk-averse agents, with risk-neutral agents being indifferent between the two solutions. In the ACAN 2011 tournament, we used a risk-seeking approach, with $r = 3$.

Having selected the time, t^*, at which the expected utility of the opponent's offers is maximised, our agent needs to choose a utility, u^*, to offer at that time. The approach that our strategy takes here is to maximise the expected utility, of making an offer of utility u at time t^*, as given by $u^* = \text{argmax}_{u \in [0,1]} E_{\text{offer}}(u, t^*)$.

The expected utility is calculated based on the probability that an offer of a given utility will be accepted. We assume that an offer of utility u will be accepted at time t^* if $u \leq u_{t^*}$. Since we have a probability distribution over u_{t^*}, we can calculate the probability that $u \leq u_{t^*}$ using the cumulative distribution $F(u; \mu_t, \sigma_t)$. Therefore, our agent's expectation for the adjusted utility (taking into consideration the risk attitude and time discounting) of an offer with utility u is given by $E_{\text{offer}}(u, t^*) = D(R(u)F(u; \mu_{t^*}, \sigma_{t^*}), t^*)$ where $D(\cdot, \cdot)$ and $R(\cdot)$ are as before, and $F(\cdot)$ is the c.d.f. for $f(\cdot)$. Due to the effect of the standard deviation, note that the risk-seeking agent will find a higher optimal utility than a risk-neutral or risk-averse agent.

Finally, having determined u^* as the utility to offer at time t^*, our agent needs to choose a target utility, u_τ to offer at the current time, t_c. The agent should not concede immediately to offer u^* at the current time, nor should it wait until t^*. Either of those approaches is likely to result in concession behaviour which is too extreme, especially since they are based on predictions which may be inaccurate. Therefore, and to avoid any additional parameters, our approach is simply to concede linearly between $[t_{\text{lr}}, u_{\text{lr}}]$ and $[t^*, u^*]$, where t_{lr} is the time at which the regression was last performed and u_{lr} is the target utility at that time. The overall concession will not generally be linear, as the predictions of t^* and u^* are revised at the end of each time window, throughout the negotiation.

4 Generating an Offer

Having selected a target utility, u_τ, our strategy needs to generate an offer which has a utility close to that target. In a multi-issue negotiation, there may be many different packages with a similar utility. Under the real-time constraints of the competition, the goal is to reach an agreement within a shorter time period, but not necessarily to limit the number of offers made.

Consequently, our agent uses a fast strategy, which simply chooses a random offer with a utility in the range $[u_\tau - 0.025, u_\tau + 0.025]$. A range is used, since, in a domain which consists entirely of discrete issues, it is highly likely that there are no possible outcomes with a utility of exactly u_τ. If an offer cannot be found even within this range, the range is expanded, until a solution is found. In addition, if the target drops below the highest value of the offers made by the opponent, we instead propose the package with that utility that was offered by the opponent. This is since we assume that, for a set of possible offers with utility greater than u_τ, the one which is most likely to be accepted is the one which has previously been offered by the opponent.

5 Conclusions and Future Work

In this work, we have developed a negotiation strategy that uses a principled approach to concession, by setting its behaviour as a best response, according to its estimate of the opponent's future concession. In future work, we plan to improve our offer generation algorithm by modelling the opponent's preferences, and then choose a package at a given utility level which maximises the likelihood that it is accepted by the opponent.

References

1. Baarslag, T., Hindriks, K., Jonker, C.M., Kraus, S., Lin, R.: The second automated negotiating agents competition (ANAC2011). In: Complex Automated Negotiations. SCI, vol. 435, pp. 183–197. Springer, Heidelberg (2012)
2. Rasmussen, C.E., Williams, C.K.I.: Gaussian Processes for Machine Learning. The MIT Press (2006)
3. Williams, C.R., Robu, V., Gerding, E.H., Jennings, N.R.: IAMhaggler: A Negotiation Agent for Complex Environments. In: Ito, T., Zhang, M., Robu, V., Fatima, S., Matsuo, T. (eds.) New Trends in Agent-Based Complex Automated Negotiations. SCI, vol. 383, pp. 151–158. Springer, Heidelberg (2012)
4. Williams, C.R., Robu, V., Gerding, E.H., Jennings, N.R.: Using gaussian processes to optimise concession in complex negotiations against unknown opponents. In: Proc. of the Twenty-Second International Joint Conference on Artificial Intelligence, vol. 1, pp. 432–438 (2011)

BRAM Agent

Radmila Fishel, Maya Bercovitch, and Ya'akov (Kobi) Gal

Abstract. BRAM considers the other agent as a partner in a negotiation process (and not as an opponent) since its strategy is to reach a solution which will satisfy the other agent, while maintaining its' own utility threshold. BRAM has four main characteristics that guide it through the tournament: simple and fast, compromises as time elapses, seeks a win-win agreement and interrupts modeling attempts by others. BRAM amounted to the 4th place in the ANAC2011 competition.

1 Introduction

BRAM was developed in the GENIUS negotiation simulator environment, and was designed for the ANAC2011 competition[1]. According to the rules of the competition, a negotiation session is conducted in a specific domain and includes two agents who interact using an alternating offer protocol. Each negotiation session is limited to 180 seconds. Once an agreement is reached, each agent receives its corresponding utility score based on the agreement. Otherwise, both agents receive a constant no-agreement score. The BRAM agent was designed to be domain-independent in order to cope with different domains and utility functions, including domains with discount factor (discount as a function of time).

BRAM's strategy is outlined as follows. First, BRAM searches for a win-win agreement that will satisfy both parties. We hypothesized that by considering the other agent not as an opponent, but as a partner with common interests, BRAM will be able to achieve better agreements in a shorter amount of time. Second, the extent

Radmila Fishel · Maya Bercovitch · Ya'akov (Kobi) Gal
Department of Information Systems Engineering at Ben-Gurion University of the Negev, Beer-Sheva, Israel
e-mail: {fishelr,lianima,kobig}@bgu.ac.il
http://mmi.tudelft.nl/negotiation/index.php/Genius

[1] http://mmi.tudelft.nl/negotiation/index.php/Automated_Negotiating_Agents_Competition_%28ANAC%29

T. Ito et al. (Eds.): Complex Automated Negotiations, SCI 435, pp. 213–216.
springerlink.com

to which BRAM is willing to compromise increases as time elapses. The decision whether to accept an offer made by the other agent, and the decision of which offer to make, are based on a utility threshold that is reduced during the negotiation. Lastly, BRAM attempts to interrupt modeling attempts by the other party. This is done by switching strategies during the negotiation session, and by including randomization in various stages of the decision process. We hypothesized that if the other agent will not be able to discern that BRAM compromises as time elapses, it will make it compromise more quickly. Finally, BRAM can generate responses to offers made by the other agent very quickly. This allows to reach agreement more quickly and benefits both parties.

2 BRAM's Strategy

At the beginning of a negotiation session, BRAM creates a bid array which contains all the possible bids in the given domain. The array creation time is limited to 2 seconds. If the set of optional bids is very large, only a random sample of these bids will be included in the array. The bid array is sorted in a descending order according to the utility values for BRAM. This bid array will be used in cases when a win-win offer was not found, in order to make a random offer.

BRAM's decision making process is presented as a flow chart in Figure 1. When BRAM receives an offer from the other agent, it first compares the utility score of this offer with its current utility threshold. If the utility to BRAM is above the threshold, it will accept the offer. Otherwise, BRAM will make an alternative offer. First, it will try to make a win-win offer which is beneficial for both parties.

In order to create win-win offers, BRAM estimates the preferences of the other agent. We assume that the more the other agent requests a given value for an issue, the more it prefers this value. Also, the other agent may change offers over time and therefore BRAM will only refer to the last 10 bids made by the other party. Based on these assumptions, BRAM creates a histogram for each issue according to the offers received from the other agent. BRAM offers a bid which gives it the highest expected score for bids that are also beneficial to the other party.

An example of the histograms of the last 10 offers made by the other agent in the Laptop domain can be viewed in Figure 2. The Laptop domain includes 3 issues, and each issue contains 3 values. It is obvious that in the *Laptop Brand* issue, the other agent clearly prefers *Dell* over *Macintosh* and *HP*. For the *External Monitor* issue, there is no clear distinction between *19"*, *20"* and *23"*, so the other agent may be indifferent between these values. Finally, in the *Hard Disk* issue, BRAM can infer that the other agent prefers the *120 GB* over *60 GB* and *80 GB*, but it may compromise.

In each round, BRAM tries to create a new offer which will give BRAM a utility score that is at least as high as its current utility threshold. If BRAM fails to create such an offer, it will use the bid array previously mentioned to make a random offer that is above its threshold. Specifically, from its last offer's position (in the array),

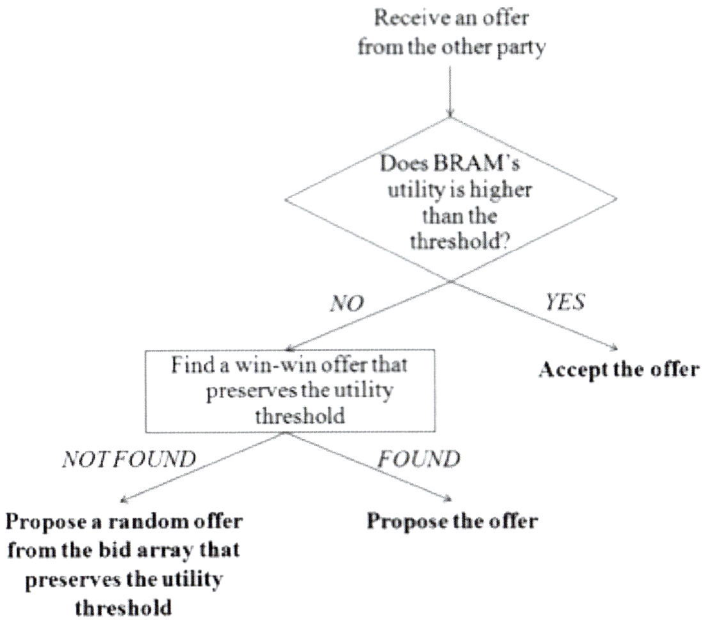

Fig. 1 Illustration of BRAM's decision making process for a single round of a negotiation process

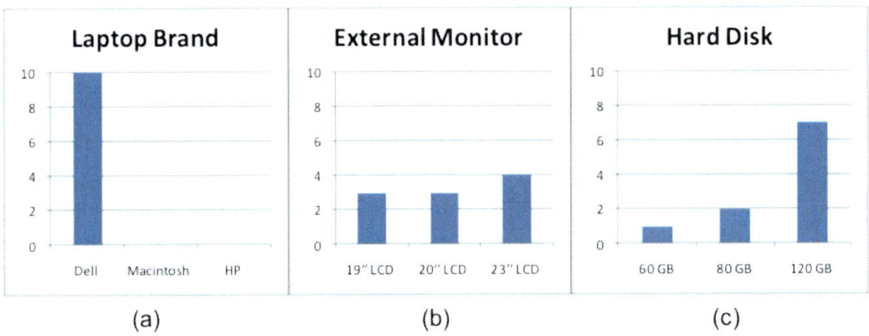

Fig. 2 An example of the histograms of the last 10 offers made by the other agent in the *Laptop* domain. Each histogram refers to an issue in this domain: (a) *Laptop Brand* issue, (b) *External Monitor* issue, and (c) *Hard Disk* issue.

BRAM chooses randomly a nearby bid from a predefined range. The range size is calculated as a function of the number of optional bids in the current domain. If the selected offer's utility is lower than the threshold value, BRAM will offer the same bid that was previously suggested by it. The number of times a bid can be offered is limited, in order to enlarge the variety of bids made by BRAM.

As mentioned, BRAM uses a utility threshold in order to decide whether to accept a given offer, or what offer to make next. The threshold is calculated as a pre-defined percentage of the maximum utility that can be achieved at the current moment and decreases as time elapses. By running empirical simulations with all of the agents in last year's competition (ANAC 2010) and our classmates who also developed agents for ANAC2011 tournament, we determined the following threshold function: during the first 60 seconds, the threshold value is 93% of the maximum utility; during the next 90 seconds it is 85% of the maximum utility; next, during the following 25 seconds it is 70%; and finally 20% (during the remaining 5 seconds).

3 Tournaments Results

In essence, BRAM is a cooperative agent, and while playing with other cooperative agents, it received quite high utilities. However, due to the fact the BRAM compromises as time elapses, playing against non-cooperative agents led to lower utilities. BRAM participated in a national tournament with agents written by students at Ben-Gurion and Bar-Ilan Universities, and received the 1st place with an average utility of 0.7705. Then, at the qualifying results of the ANAC2011 competition, BRAM received a utility of 0.6904 and finished in the 6th place. Because the distance from the 1st place was very small, we decided (based on empirical simulations) to toughen BRAM's approach and set the utilities thresholds to higher percentage values. In the final competition, BRAM won the well respected 4th place in ANAC2011 by receiving an average utility of 0.6746.

A posthoc analysis of the results revealed that in many of its agreements, BRAM and its partners reached the Nash-Equilibrium point. Moreover, BRAM failed to reach agreement with its partner in just 2 out of 385 negotiation sessions (~0.005%). These results were consistent across domains, further demonstrating the efficacy of our negotiation strategy.

TheNegotiator: A Dynamic Strategy for Bilateral Negotiations with Time-Based Discounts

A.S.Y. Dirkzwager, M.J.C. Hendrikx, and J.R. De Ruiter

Abstract. Recently developed automated negotiation agents are starting to outperform humans in multiple types of negotiation. There has been a large body of research focusing on the design of negotiation strategies. However, only few authors have addressed the challenge of time-based discounts. In the ANAC2011, negotiation agents had to compete on various domains both with and without time-based discounts. This work presents the strategy of one of the finalists: *TheNegotiator*. Our contribution to the field of bilateral negotiation is threefold; First, we present the negotiation strategy of *TheNegotiator*; Second, we analyze the strategy using various quality measures; Finally, we discuss how the agent could be improved.

1 Introduction

Past decade, virtual negotiation agents were developed which outperformed humans on some negotiation domains [3]. There is an increasing interest in using automated negotiation in e-commerce settings [5]. This interest is fueled by the promise of computer agents outperforming human negotiators [5], reducing the time and negotiation costs [2], and the ability to improve the negotiation skills of humans [4].

Despite the potential of automated negotiations within the e-commerce domain, there has been limited development in applying the theory to practice [5]. In addition, while many negotiation strategies have been developed, only few take the realistic setting of time-based discounts into account.

Towards stimulating the development of negotiation strategies for environments with time-based discounts, this work discusses *TheNegotiator*. Our contribution to

A.S.Y. Dirkzwager · M. Hendrikx · J.R. De Ruiter
Man-Machine Interaction Group, Delft University of Technology, Mekelweg 4, Delft, The Netherlands
e-mail: {A.S.Y.Dirkzwager,M.J.C.Hendrikx}@student.tudelft.nl
J.R.deRuiter@student.tudelft.nl

T. Ito et al. (Eds.): Complex Automated Negotiations, SCI 435, pp. 217–221.
springerlink.com

the field of bilateral negotiation is threefold; First, Section 2 discusses the negotiation strategy; Second, Section 3 uses a toolkit of quality measures implemented in Genius to analyze the negotiation strategy using various quality measures; Finally, Section 3 provides directions for future research.

2 Negotiation Strategy

TheNegotiator uses a time-based, domain-dependent strategy. Section 2.1 specifies the concession curve. Section 2.2 discusses the acceptance strategy. Section 2.3 describes the bidding strategy.

2.1 Derivation of the Concession Curve

The concession curve of *TheNegotiator* depends on the mode – discount or no discount (a number between 0 and 1) – and the different phases. Initially, the negotiation starts in no discount mode. A transition to discount mode occurs after the negotiation time surpasses the discount factor. For a negotiation with no discount factor, this entails that a transition does not occur as the discount is 1. Each of the two modes has the same three phases. In each phase a different bidding and acceptance strategy is used to increase the acceptation chance. The time of phase transition is determined differently for both modes.

In the no discount mode, *TheNegotiator* creates three phases based on the duration of the negotiation: the length of the first phase is $\frac{28}{36}$, second phase $\frac{7}{36}$, and the final phase $\frac{1}{36}$ of the time. The next step is to divide the utility range into three ranges [0 - $\frac{5}{8}$], [$\frac{5}{8}$ - $\frac{7}{8}$], and [$\frac{7}{8}$ - 1]. In each phase a linear concession curve determines the target utility. The target utility of the first phase starts at the maximum utility and decays linearly to $\frac{7}{8}$ of the full utility range. Given the length of the phase, this entails that *TheNegotiator* concedes very slowly during most part of the negotiation. Next, the second phase starts with a utility of $\frac{7}{8}$ and decays to $\frac{5}{8}$. Finally, the last phase considers the remaining range.

In the discount mode, which starts when discount time has passed, the remaining time and utility range is determined. To concede more quickly, each phase is set to visit $\frac{1}{3}$ of the remaining utility range. The duration of each phase is determined by dividing the number of bids in the range by the total amount of bids and multiplying by the remaining time. In contrast to the no discount mode, an equal amount of time is spend on each bid. Given the utility range and time per phase, the target utility can be found by using linear interpolation identical to the no discount mode.

2.2 Acceptance Strategy

TheNegotiator uses the concession curve to determine if a bid should be accepted. In all phases, a bid is accepted when the utility of the bid is higher or equal to the target utility. In the third phase there is the extra condition that a bid is accepted without further consideration if *TheNegotiator* estimates that there are less than 15 player moves remaining, where a move is defined as an action by both agents. The amount of rounds left is determined by using Equation 1, where μ_{15} average time between the last 15 bids and $(T-t)$ is the remaining time.

$$RoundsLeft = \frac{(T-t)}{\mu_{15}} \tag{1}$$

2.3 Bidding Strategy

TheNegotiator uses the concession curve to offer bids. A random bid with a utility within the target utility window $[l, u]$, which is calculated by using the concession curve, is chosen to offer. The lowerbound l of the window corresponds to the target utility calculated by using the concession curve. The upperbound u is determined by using Equation 2.

$$u = l + 0.2 \cdot (1 - l)] \tag{2}$$

In the first phase, 70% of the time a random bid is chosen with a utility within this range. In the other cases, the upperbound u is ignored. In the second and third phase, *TheNegotiator* uses opponent bids which are above the lowerbound l A random bid is chosen from these bids as a counteroffer. If there is not a single bid which falls in the target utility window, then the same strategy as the first phase is applied.

3 Empirical Evaluation

This Section evaluates the negotiation strategy of *TheNegotiator*. Section 3.1 discusses the setup of the tournament used to evaluate the quality of the negotiation strategy. Section 3.2 discusses which quality measures are used to evaluate the negotiation strategy. Section 3.3 evaluates the results.

3.1 Tournament Setup

From the eight domains of the ANAC2011, five domains were chosen based on size and opposition (Adg, Camera, Grocery, IS_BT_Acquisition, and Laptop). Large domains and discounts were not used, as most ANAC2010 agents do not handle these well.

The top three agents of the ANAC2010 (Agent K, Yushu, Nozomi), ANAC2011 (HardHeaded, Gahboninho, IAMhaggler2011), and *TheNegotiator* competed on these five domains in order to establish the quality of the negotiation strategy of *TheNegotiator*. The tournament results were averaged over three runs to increase the reliability of the results and four computers were used in combination with a distributed version of Genius which divided the 1680 sessions.

3.2 Quality Measures

Three types of quality measures were implemented in Genius to evaluate the quality of the negotiation strategy: tournament measures (average time of agreement, percentage of agreement, and average utility), outcome quality measures (average Pareto distance, average Nash Distance) and trajectory analysis (percentage of unfortunate moves) [1]. The results are calculated after a negotiation session to ensure that the calculation does not influence the session.

3.3 Evaluation of Results

The results of the tournament are shown in Table 1. The average time of agreement was high, this is in line with the winners of the ANAC2010 (Agent_K) and ANAC2011 (HardHeaded). Despite not having an opponent model, *TheNegotiator* scores above average on unfortunate moves. In many of the cases (averaged final utility, averaged Pareto distance, averaged Nash distance and percentage of agreement) the results from the *TheNegotiator* differed only slightly from the other agents.

The avg. time of agreement gives us reason to believe that the acceptance criteria could be improved to increase the avg. utility. We believe that the avg. utility could

Table 1 Overview of the tournament results. The best value is bold; the worst underlined.

Agent	Avg. time of agreement	Percentage agreement	Avg. utility	Avg. Pareto distance	Avg. Nash distance	Percentage unfortunate
Agent_K (2010)	0.651	99.167	0.881	0.012	0.106	0.108
Gahboninho (2011)	0.563	99.167	0.821	0.012	0.116	<u>0.158</u>
HardHeaded (2011)	0.673	<u>98.056</u>	0.891	<u>0.024</u>	<u>0.148</u>	**0.034**
IAMhaggler 2011 (2011)	**0.325**	98.333	<u>0.794</u>	0.022	0.126	0.142
Nozomi (2010)	0.583	**99.722**	0.887	**0.007**	**0.085**	0.042
TheNegotiator (2011)	0.626	99.167	0.873	0.013	0.107	0.064
Yushu (2010)	<u>0.681</u>	99.167	**0.928**	0.012	0.110	0.071

be even further improved by adding a frequency model similar to HardHeaded to decrease the percentage of unfortunate moves.

It should be noted that the results are not in line with the ANAC competition results. This can partly attributed to the selection of domains – there is no large domain in the set of domains as ANAC2010 agents do not scale well – and the usage of only domains without discounts.

4 Conclusion and Future Work

In this paper, we presented an negotiation strategy which allows an agent to negotiate efficiently in domains with time-based discounts. We provided an overview of the strategy of *TheNegotiator* (see Section 2) which is domain dependent. The strategy of *TheNegotiator* was empirically evaluated against the best ANAC2010 and ANAC2011 agents (see Section 3) using different types of quality measures. For future work, we plan to adapt the negotiation strategy such that it takes the opponent's behavior into account by employing a Bayesian or Frequency model.

Acknowledgements. We would like to thank Catholijn Jonker, Koen Hindriks, and Tim Baarslag for introducing us to the field of bilateral negotiations. Furthermore, we thank the Universiteitsfonds Delft and the Man-Machine Interaction Group of the Delft University of Technology for sponsoring our trip to the AAMAS2011.

References

1. Bosse, T., Jonker, C.: Human vs. computer behavior in multi-issue negotiation. In: Rational, Robust, and Secure Negotiation Mechanisms in Multi-Agent Systems, pp. 11–24. IEEE (2005)
2. Carbonneau, R., Kersten, G., Vahidov: Predicting opponent's moves in electronic negotiations using neural networks. Expert Systems with Applications 34, 1266–1273 (2008)
3. Lin, R., Kraus, S., Wilkenfeld, J., Barry, J.: Negotiating with bounded rational agents in environments with incomplete information using an automated agent. Artificial Intelligence 172, 823–851 (2008)
4. Lin, R., Oshrat, Y., Kraus, S.: Investigating the benefits of automated negotiations in enhancing people's negotiation skills. In: Proceedings of The 8th International Conference on Autonomous Agents and Multiagent Systems (2009)
5. Lomuscio, A.R., Wooldridge, M., Jennings, N.R.: A classification scheme for negotiation in electronic commerce. Group Decision and Negotiation 12, 31–56 (2003)

HardHeaded

Thijs van Krimpen, Daphne Looije, and Siamak Hajizadeh

Abstract. In this paper the strategy of Hardheaded negotiating agent is described. Our agent won the Automated Negotiating Agents Competition 2011. As the name implies, the agent is hardheaded, it will not concede until the very end. Useing a concession function, it generates bids in a monotonic way, which resets to a random value after the dynamic concession limit is reached. In practice, this means that most of the time the agent will cycle through the same range of bids. Since the preferences of the opponent are not known, the agent tries to learn the opponent's preference profile. It chooses bids which it thinks are optimal for the opponent in case there are equivalent bids for itself.

1 Introduction

In the competitions, agents perform bilateral multi-issue closed negotiation. That is the agent does not know what the preferences of its opponent are. Hardheaded uses a concession function that over time progressively concedes. Furthermore it is equipped with a learning module that analyzes the bids that are made by the opponent and based on this information assumes that the opponent has a certain preference. Using this knowledge, the agent tries to choose the best bid for its opponent from a set of bids that are equivalent for itself. In this paper the general strategy and particularly the learning function will be described in details. We will conclude with a discussion the agent's performance.

2 Overall Structure

This section describes the structure of our agent. First we describe the general strategy, how the agent decides which bid it will make, and how it gradually concedes

---

Thijs van Krimpen · Daphne Looije · Siamak Hajizadeh
Delft University of Technology
e-mail: {T.M.vanKrimpen,D.Looije,S.Hajizadeh}@student.tudelft.nl

T. Ito et al. (Eds.): Complex Automated Negotiations, SCI 435, pp. 223–227.
springerlink.com

the minimum utility of accepting bids. Then the learning function is described in more detail. Finally the choice of the parameters is motivated.

2.1 General Strategy

At the beginning of a negotiation session, the agent starts with computing all possible bids and their utility for itself. These are stored in a search tree for rapid retrieval. In a normal domain, the agent uses a Boulware function during the negotiation, to calculate the utility P to which the agent is willing to concede at that moment. A concession function F from [1] determines the concession limit.

$$P = \text{MinUtility} + (1 - F(t)) \cdot (\text{MaxUtility} - \text{MinUtility}) \tag{1}$$

$$F(t) = k + (1 - k)\left(\frac{\min(t, T)}{T}\right)^{\frac{1}{e}} \tag{2}$$

Here t is the current time as a proportion of total negotiation time T and k multiplied by the size of the interval [MinUtility − MaxUtility], determines the initial minimum acceptable utility. On discounted domains, the discount factor is ignored when it is greater than 0.8, otherwise the agent changes e to a value higher than 1. As on a scale from 0 to 1, time passes ahead of the discount value. The effect is that the agent suddenly changes strategy and concedes rapidly to prevent further loss of utility. The agent keeps a threshold MinUtility below which no offers are accepted. Bids that are generated are all in a narrow range deemed equivalent for itself. This range will increase monotonically towards the concession limit defined by $P(t)$, which when reached causes the agent to reset the range to a random value greater than the concession limit. In case the learning function performs inadequately, the agent also remembers the best bid that the opponent has generated so far. HardHeaded will give preference to this bid over the ones generated by the agents own learning module, provided that the opponents best bid is above the current reservation value.

2.2 Learning Module

HardHeaded negotiates in domains containing multiple issues where each issue has multiple values. The utility for each value is assumed to be unknown. Furthermore the values are unordered. The utility for a particular bid b is denoted by U_b and is the weighted sum of each of the selected values in b.

$$U_b = \sum_{n=1}^{N} w_n \times v_{n,i} \tag{3}$$

Here w_n is the weight for issue n and $v_{n,i}$ is the utility of value i that is assigned by b to issue n. The goal of the learning module is to learn the weights and utillity value for the opposing agent. To do this the agent makes two implicit assumptions regarding the opponent. Firstly, it assumes that the opponent restricts the bids it makes to a, possibly moving, limited utility range. Secondly, opponent prefers to explore the acceptable range rather than simply offering the same bid over and over again. Our learning function is a greedy reinforcement learning function, which updates the issue weights and value utilities of the preference profile after each bid. The update rule is split into 2 parts, one for the weights the other for the issue value utilities.

Listing 6. update rule

if $T == 1$ **then**
  Initialize preference profile
else
  set *index* to where $V(T) == V(T-1)$
  $W(index) = W(index) + \varepsilon$
  $W = W / \sum_{n=1}^{N} w_n$
  V(index)=V(index)+1
end if

The algorithm checks for issue values that the opponent has kept unchanged over all its offered bids. It then adds a value ε to each of those issue weights and the weights are normalized every round. The values get their utilities incremented each time they remain unchanged. Since these utilities are not required to be percentages, they only normalized to their maximum value per issue at the time the utility is calculated. Figure 1 shows a testing of the learning algorithm against an agent which uses an exponential concession function. The agent draws a bid from a range of 300 closest bids above the concession value and offers 1500 bids in total. The utility of all the possible bids which can be generated from the preference profile, is compared to the utility estimated for the same bids by the learning module using $\varepsilon = 0.2$. We have run the above scenario 100 times.

The learning function always manages to identify the most valuable bid, and the least valuable bid for the opponent. However it generally undervalues the opponents bid's utility by roughly 0.2. This offset is not important provided that the order is correct. The average standard deviation of the estimated utility is 0.13, but the mode 0.065. The standard deviation diminishes towards the extremities of the utillity space i.e. in the neighborhood of utility 1 and 0, as is demonstrated in figure1. It also illustrates how the function tends to overvalue certain bid configurations causing a gap or a spread. The overall order of estimated utilities matches the actual opponent's utility function well in this case. There are situations however that this fine auto-adjustment does not occur. Especially when the opponent follows a non-monotone pattern of concession and bid offering, the learning module can lose track of the most important issues and issue values. This can also happen when the opponent has a utility profile which has multiple equally important issues.

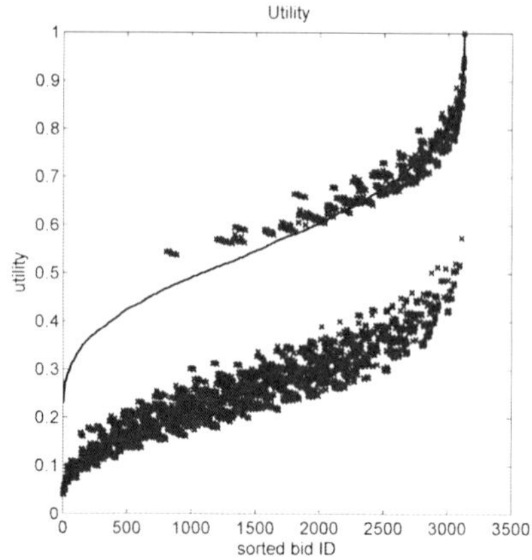

Fig. 1 The solid line shows the utility as generated by the preference profile, the dark x's show the learned utillity.

2.3 *Setting Parameters*

The agent's performance is dependent on several factors, the domain on which it plays, the opponents, and the computer it runs on. As such the parameter values can only be fine-tuned to a certain extent without losing generality. The parameters which needed tuning were the initial concession k, the discount rate coefficient e, and minimum acceptable utility $MinUtility$. The agent was tuned on domains and against agents from previous ANAC competitions, on three domains selected in one case for it sheer size (Itex vs. Cypress), in another for its zero-sum like preference profiles (inheritance domain), and the third as a generic case (car domain). The agent was tuned against three other agents, agent K, negotiator agent, and IAmHaggler. It was discovered that against agent K a large initial concession increased its willingness to concede later in the negotiating session. Due to this the initial concession k was set at 0.05 or such that at least 5 bids can be made.

Tuning $MinUtility$ and e is a balancing act between holding out as long as possible to get the best negotiation result, and conceding rapidly enough to prevent the agent from timing out. For the ANAC competitions we set $\min_{util}$ value of 0.58, and a concession rate coefficient of 0.02 by only testing our agent against other available agents and in several available domain. This values however are ad hoc and may not result the same achievements in different domains and settings.

2.4 *Discounted Domains*

The discount factor undermines the overall strategy because the general strategy is to hold out as long as possible. The discount factor makes it attractive to reach an agreement as soon as possible. However the constraint is still that if an agent concedes too rapidly to reach a quick agreement then that agent will be taken advantage of by agents who don't share their opponent's temporal preference. The solution presented here was inspired again by Fatima *et al* [1]. As mentioned earlier, The agent creates a virtual deadline at $T_{max} \times \delta$ where δ indicates the discount factor. Also a dynamic minimum utility threshold is calculated according to equation 4.

$$DynamicMinUtility = (MaxUtility - MinUtility) \times \delta + MinUtility \tag{4}$$

3 The Competition

HardHeaded won ANAC2011. Oddly enough, HardHeaded performed better in the final rounds than in the qualifying rounds. There were few notable things in the results of the qualifying rounds. Most important was the number of time-outs, overall about 33%. But when only looking at the results of the qualifying rounds against the other finalists, about 25% of the negotiations ended with a time-out. Furthermore, the average utility reached was higher against only finalists than against all participants in the qualifying rounds. This could explain the better performance in the final rounds.

4 Conclusion

The agent has a low computational complexity, this enables it to generate bids very rapidly. This celerity allows the agent to concede quickly when the time of the negotiating session is about to finish, whilst still thoroughly exploring the bid space. It uses a simple learning module and an optimal concession function to generate bids. The agent does not need a more sophisticated learning algorithm because it has ample time to explore the bid-space and offer any bid. An improvement can be done by estimating the opponent's rate of concession for discounted domains and switching strategies in accord.

Reference

1. Fatima, S.S., Wooldridge, M., Jennings, N.R.: Optimal Negotiation Strategies for Agents with Incomplete Information. In: Meyer, J.-J.C., Tambe, M. (eds.) ATAL 2001. LNCS (LNAI), vol. 2333, pp. 377–392. Springer, Heidelberg (2002)

A Tit for Tat Negotiation Strategy for Real-Time Bilateral Negotiations

Tim Baarslag, Koen Hindriks, and Catholijn Jonker

Abstract. We describe the strategy of our negotiating agent, Nice Tit for Tat Agent, which reached the finals of the 2011 Automated Negotiating Agent Competition. It uses a Tit for Tat strategy to select its offers in a negotiation, i.e.: initially it cooperates with its opponent, and in the following rounds of negotiation, it responds in kind to the opponent's actions. We give an overview of how to implement such a Tit for Tat strategy and discuss its merits in the setting of closed bilateral negotiation.

1 Introduction

This paper presents a new negotiation strategy, called the Nice Tit for Tat Agent, which we developed and entered into the Second Automated Negotiating Agent Competition[1] (ANAC2011). ANAC is a tournament between a set of negotiating agents which perform closed bilateral negotiation using the alternating offers protocol. The negotiation environment consists of multi-issue scenarios, and is closed in the sense that there is uncertainty about the opponent's preferences.

Our negotiation strategy is based on the principle of Tit for Tat: cooperating on the first move and then mirroring whatever the other player did in the preceding round. Thus, Tit for Tat is a strategy of cooperation based on reciprocity [1].

Tit for Tat has been applied and found successful in many other games, including the Iterated Prisoner's Dilemma game. It is considered to be a very robust strategy, mainly because of the following three features:

i. It is never the first to defect (i.e., it plays nice as long as the the opponent plays nice as well);

Tim Baarslag · Koen V. Hindriks · Catholijn M. Jonker
Delft University of Technology, The Netherlands
e-mail: {T.Baarslag,K.V.Hindriks,C.M.Jonker}@tudelft.nl

[1] http://mmi.tudelft.nl/anac

T. Ito et al. (Eds.): Complex Automated Negotiations, SCI 435, pp. 229–233.
springerlink.com

ii. It can be provoked into retaliation by a defection of the opponent;
iii. However, it is forgiving after just one act of retaliation.

In this paper, we discuss how to implement such a Tit for Tat strategy in the setting of closed bilateral negotiation.

2 A Tit for Tat Negotiation Strategy

We first give a broad overview of the Tit for Tat negotiation strategy. The ideas behind and details of implementation are given below.

The Nice Tit for Tat agent plays a tit for tat strategy with respect to its own utility. The agent will initially cooperate, then respond in kind to the opponent's previous action, while aiming for the Nash point of the negotiation scenario. After each move by the opponent, it updates its Bayesian opponent model to make sure it responds with a nice move to a concession by the opponent.

2.1 Four Ways of Repaying the Favor

When implementing a Tit for Tat strategy, the first question that needs answering is: in what way should an agent reciprocate? In closed bilateral negotiation, there are two utility functions in play, one of which is unknown to the other agent. Therefore, there are two different actions from the opponent that could be considered 'nice':

i. The opponent concedes according to its own utility function;
ii. The opponent offers more utility to the agent, according to the agent's utility function.

Both actions can be reciprocated by the agent by again choosing one of the two options. Therefore, there are four different ways for a Tit for Tat agent to reciprocate (see Fig. 1).

For the Tit For Tat agent, we have elected to reciprocate according to the agent's own utility function. That is, when the opponent makes a bid that is better for the agent, in return, the agent will produce a bid of lower utility for itself. There is a good reason to do so: the agent's own utility function is known to the agent. The other three ways to reciprocate depend on the utility information provided by an opponent model, which is inherently unreliable.

2.2 Reciprocating and Making Nice Moves

The basic idea of the Tit for Tat strategy is to reciprocate in terms of the agent's own utility as described in Sec 2.1, but it also takes into account the utility of its opponent. To do so, it constructs an opponent model using Bayesian learning [4, 5].

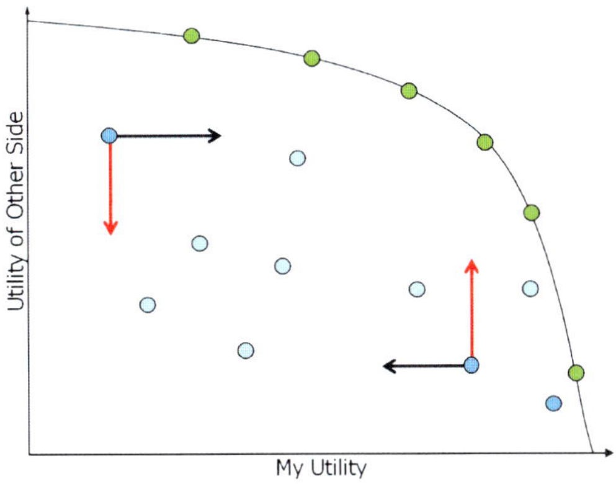

Fig. 1 Four different ways to reciprocate

Using this opponent model, a naive implementation of the Tit for Tat strategy would go as follows:

i. Measure the opponent's concession in terms of the agent's own utility function;
ii. Mirror this bid as described in Sec 2.1, sacrificing the same amount as is offered by the opponent;
iii. Make the offer as attractive as possible for the opponent using the Bayesian opponent model.

However, this implementation would lead to an agent strategy that is far too nice: if the opponent yields $\frac{1}{2}$ utility to the agent, then it would respond in kind by sacrificing the same amount. In other words: the agent would be satisfied with deals of $\frac{1}{2}$ utility. Given that many domains of ANAC have win–win outcomes with utilities much higher than that, this is clearly a suboptimal approach.

Therefore, in our version of the agent, it generally aims for more than half, depending on the negotiation scenario. Instead, it uses the opponent model to make an estimate of the location of the *Nash point* of the negotiation scenario, and then aims for this outcome (see Fig. 2). For example: if the opponent has made an offer that is 70% on the way to the Nash point, the agent will respond in kind by approaching from the other side, making an offer that is 30% away from the Nash point.

2.3 Acceptance Strategy

The preceding sections focused on the bidding strategy of the Tit for Tat agent. However, there is a second component of a negotiator's strategy that can also highly influence the outcome of a negotiation, namely its acceptance strategy. In [2], several acceptance conditions are defined that are designed to perform well in conjunction

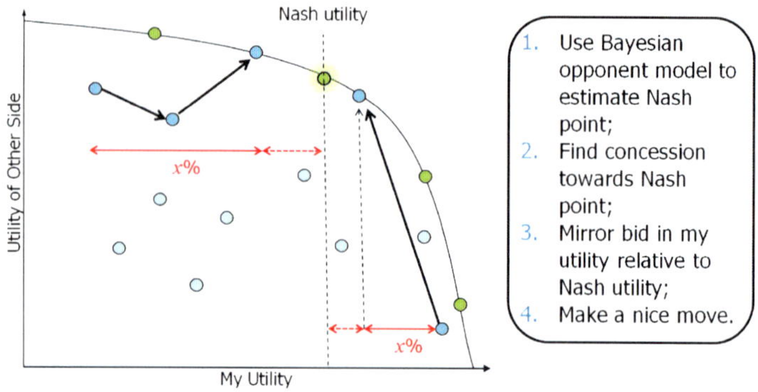

Fig. 2 The four steps taken by the Tit for Tat bidding strategy to determine the next offer

with an arbitrary bidding strategy. The Nice Tit for Tat agent is equipped with a particular type of acceptance condition called $\mathbf{AC}_{\text{combi}}$, which the paper shows to work better than the majority of more simple generic conditions.

The basic idea behind $\mathbf{AC}_{\text{combi}}$ is as follows: in case the bidding strategy plans to propose a deal that is worse than the opponent's offer, it has reached a consensus with the opponent and thus accepts the offer. However, if there still exists a gap between the two offers and time is short, $\mathbf{AC}_{\text{combi}}$ waits for an offer that is not expected to improve in the remaining time.

3 Conclusion and Future Work

In this paper, we have provided an overview of the strategy of the Nice Tit for Tat Agent, which has participated in the finals of the ANAC2011 competition.

We have designed a new negotiation strategy based on the well-known Tit for Tat principle. We have covered both the procedure that our agent follows to select a bid that reciprocates the play by its opponent, and the way it chooses to accept a certain negotiation outcome.

As has been found in post-tournament analysis of the ANAC results [3], the Tit for Tat agent was the only agent in the tournament to match the behavior of the opponent (which is to be expected). This means it plays tough against hardheaded negotiators, but it also means it plays nice against strategies that concede easily. It is observed in the paper that this approach might not be as successful in negotiation as in some other games, because it does not exploit the conceding strategies enough to reach the top rankings. More research is required to find this delicate balance between cooperative and competitive behavior of a negotiating agent.

References

1. Axelrod, R.: The Evolution of Cooperation. Basic Books (1984)
2. Baarslag, T., Hindriks, K., Jonker, C.: Acceptance conditions in automated negotiation. In: Proceedings of The Fourth International Workshop on Agent-based Complex Automated Negotiations, ACAN 2011 (2011)
3. Baarslag, T., Hindriks, K., Jonker, C.: Towards a Quantitative Concession-Based Classification Method of Negotiation Strategies. In: Kinny, D., Hsu, J.Y.-J., Governatori, G., Ghose, A.K. (eds.) PRIMA 2011. LNCS, vol. 7047, pp. 143–158. Springer, Heidelberg (2011)
4. Hindriks, K.V., Tykhonov, D.: Opponent modelling in automated multi-issue negotiation using bayesian learning (2008)
5. Zeng, D., Sycara, K.: Bayesian learning in negotiation. International Journal of Human Computer Systems 48, 125–141 (1998)

AgentK2: Compromising Strategy Based on Estimated Maximum Utility for Automated Negotiating Agents

Shogo Kawaguchi, Katsuhide Fujita, and Takayuki Ito

1 An Implementation of Negotiating Agents Based on Compromising Strategy

1.1 Other Agent's Analysis and Basic Strategy

In the setting of the competition, my utility space is not mutually taught the other agent. Therefore, there is a little information can be used for the strategy construction. In this thesis, it proposes the technique for expecting the best mutual agreement offer that can be get out from the other agent in the future from the statistical information of the value in which the other agent's proposal is evaluated it's own utility space. Moreover, it is assumed that the agent's behavior is decided to compromise to the best mutual agreement offer. Concretely, agent's own behavior is decided based on the following expressions (1) and expressions (2).

$$emax(t) = \mu(t) + (1 - \mu(t))d(t) \tag{1}$$

$$target(t) = 1 - (1 - emax(t))t^{\alpha} \tag{2}$$

$emax(t)$ shows the maximum utility value presumed to t which is receipt from the other agent. It calculates by average $\mu(t)$ of value in which offer of other agent who accumulates and collected is evaluated by own utility space and width $d(t)$ of action until time t. The width of the behavior of the other agent is presumed by deflection,

Shogo Kawaguchi · Takayuki Ito
Nagoya Institute of Technology
e-mail: kawaguchi@itolab.mta.nitech.ac.jp, ito.takayuki@nitech.ac.jp

Katsuhide Fujita
Nagoya Institute of Technology / Massachusetts Institute of Technology
e-mail: fujita@itolab.mta.nitech.ac.jp

T. Ito et al. (Eds.): Complex Automated Negotiations, SCI 435, pp. 235–241.
springerlink.com

and whether a how much advantageous offer for me can be get out from the other agent by tempering with the average is considered. The width $d(t)$ of the action until time t is calculated by decentralization of value in which offer of other agent who accumulates and collected is evaluated by own utility space.

When assuming that other agent's offer was generated in own utility space based on uniform distribution. Decentralization can be calculated as follows.

$$\sigma^2(t) = \frac{1}{n}\sum_{i=0}^{n} x_i^2 - \mu^2 = \frac{d^2(t)}{12} \quad (3)$$

Therefore, when you calculate the width $d(t)$ of the action from decentralization

$$d(t) = \sqrt{12}\sigma(t) \quad (4)$$

The width $d(t)$ of the action of the other agent is presumed from decentralization. When the mean value of the action is located at the center of the domain of the utility value, it can be thought that the maximum value that can be get out from the other agent is harmony of $1/2$ of the average and the width of the action. However, it is possible to move only in the direction where the utility value is high when the average of the utility value in the utility space of other agent's offer is low, and when the average is high oppositely, the action can be expanded only in a low direction. Therefore, accurate presumption is done by assuming the mean value to be weight. $target(t)$ is a standard of the utility value of the proposal at time t, and α is a coefficient in which the speed of the compromise is adjusted. Ending the negotiation early doesn't have the advantage in this rule. Therefore, it is effective that it searches for other agent's trend consuming the time limit, and repeating the proposal mutually, and both search for both mutual agreement candidate ideas with high utility value. However, because it is a tournament form, own utility value is requested to be raised as much as possible. Therefore, it proposes the offer that my utility value rises, and it proposes the action that approaches the maximum value that can be get out from the other agent presumed with the time passage immediately after beginning of the negotiation.

Figure 1 is an example of $target(t)$ that changes α when $emax(t)$ is set by $\mu(t) = \frac{1}{10}t \quad d(t) = \frac{2}{5}t^2$ from 1 to 9.

1.2 Control of Approach

It is not possible to correspond when the other agent take a hard stance for a simple strategy. It concede that the other agent in the hard stance too much when approaching in the maximum.

Figure 2 is an example of $target(t)$ that changes α when $emax(t)$ is set by $\mu(t) = 0 \quad d(t) = \frac{1}{10}$ from 1 to 9. When the approach is not controlled, it concedes to the other party according to $target(t)$ without bounds as shown in figure 2.

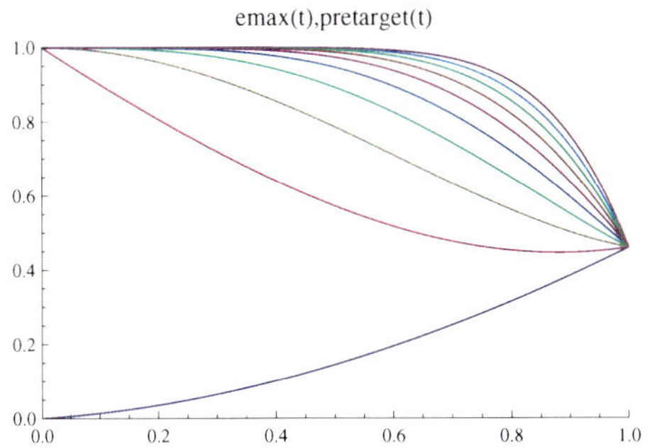

Fig. 1 Example of $target(t)$ when $emax(t)$ is set by $\mu(t) = \frac{1}{10}t \quad d(t) = \frac{2}{5}t^2$

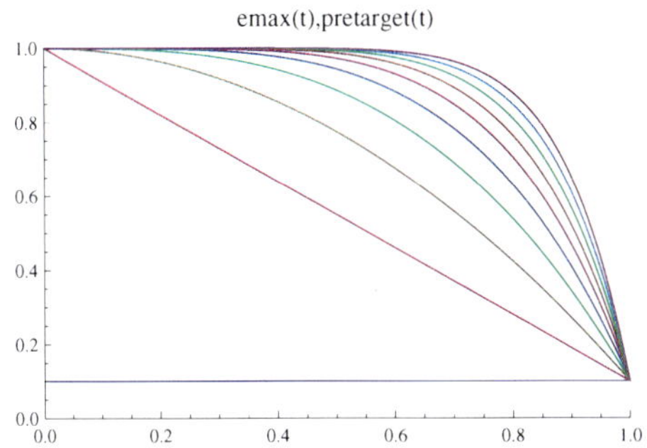

Fig. 2 Example of $target(t)$ when $emax(t)$ is set by $\mu(t) = 0 \quad d(t) = \frac{1}{10}$

Then, the the degree of concession of each is measured. When own concession degree is too larger than the width of the presumed behavior of the other agent, the approach is slowed down. The degree of a concession each other is presumed by the following expressions (5). The degree of the lowest concession at time t is assumed to be $g(t)$.

$$ratio(t) = \begin{cases} \frac{d(t)+g(t)}{1-target(t)} & \text{if } \frac{d(t)+g(t)}{1-target(t)} < 2 \\ 2 & \text{otherwise} \end{cases} \quad (5)$$

The movement of $ratio(t)$ when $emax(t)$ is set by $\mu(t) = 0 \quad d(t) = \frac{1}{10}$ is figure 3.

When I am conceding too much compared with the other party, $ratio(t)$ approaches 0 as shown in figure 3. Therefore, an excessive approach is controlled by assuming concession degree $ratio(t)$ to be weight of the target of the action. $target(t)$ is defined by using $ratio(t)$ as expression (6).

$$target(t) = ratio(t) * (1 - (1 - emax(t))t^{\alpha}) + (1 - ratio(t)) \quad (6)$$

When the other agent is amiable, it is possible to concede to the other party quickly by tempering with the degree of a concession each other and controlling own

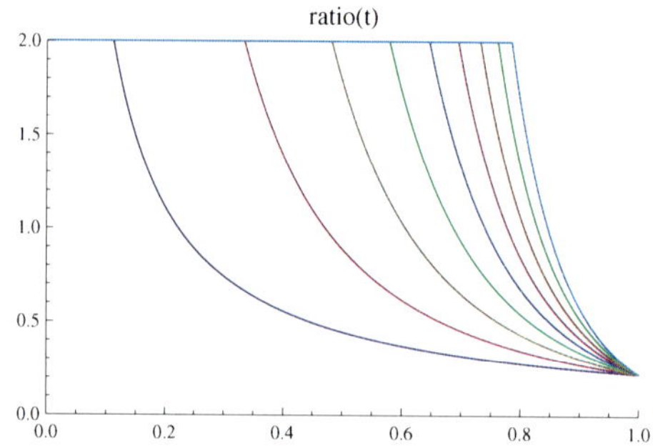

Fig. 3 Example of $ratio(t)$ when $emax(t)$ is set by $\mu(t) = 0 \quad d(t) = \frac{1}{10}$

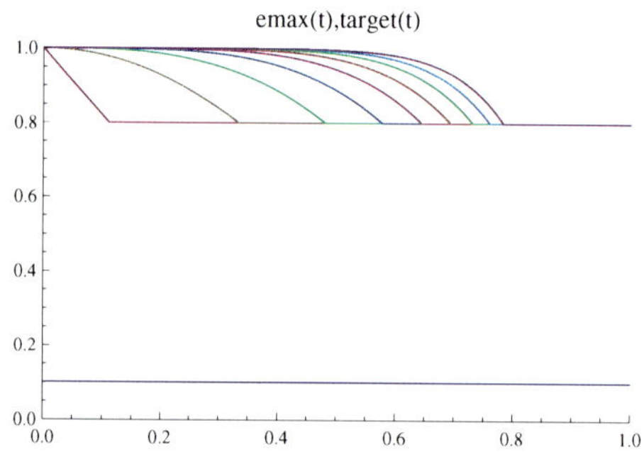

Fig. 4 $target(t)$ that introduced $ratio$

behavior. It becomes possible to model the action that doesn't concede more than constancy when opposing it oppositely.

The appearance of $target(t)$ that introduces $ratio(t)$ is figure 4. The thing that an excessive approach is controlled compared with figure 2 can be perceived in figure 4.

1.3 Coefficient Selection

Next, the function that selects coefficient α is shown. It had swinging at the first stage of the negotiation, and a steady action was designed in the last stage. Therefore, the random nature is introduced into the compromise coefficient. However, to use *target* to judge whether to accept the other agent's offer, the judgment should make sure there is nothing swing. Therefore, it is time when it makes own offer in case of occasion to judge other agent's offer and a different coefficient is used. Of each is assumed to be $\alpha(t)$ and $\beta(t)$, and the width of swinging of the action is decided as τ by the following expressions (7) and (8).

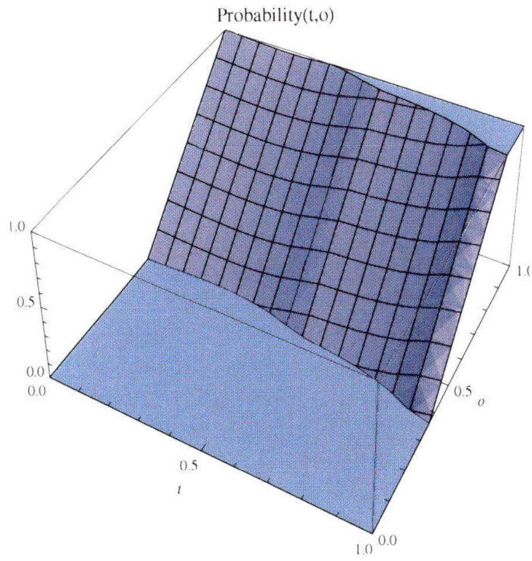

Fig. 5 Acceptance probability space

$$\alpha(t) = 1 + \tau + 10\mu(t) - 2\tau\mu(t) \tag{7}$$

$$\beta(t) = \alpha + random[0,1] * \tau - \frac{\tau}{2} \tag{8}$$

Expression (7) is a coefficient used when whether the other agent's offer is accepted is judged. Expression (8) is a coefficient used when own proposal is made. $random[0,1]$ of expression (8) generates a random value between from 0 to 1.

1.4 Decision of My Offer, and Evaluation of Other Agent's Offer

The idea which offer to select in my utility space was shown. It searches for the offer that should be proposed based on it. The search becomes difficult as the issue increases. Then, it searches for alternatives with the utility value of $target(t)$ while changing the starting position of the search at random by using iterative deepening depth-first search.

Next, evaluation whether to accept other agent's offer. Whether it accepts is judged stochastically based on the distance with target $target(t)$ and the distance from the average. The expression of the probabilistic computation of the acceptance is (9).

$$P = \frac{t^5}{5} + (\text{Offer} - emax(t)) + (\text{Offer} - target(t)) \tag{9}$$

Acceptance probability P is calculated by using time passage, value Offer in which offer by other agent is evaluated by own utility space, $target(t)$ when $\alpha(t)$ is used for coefficient of approach, and estimated maximum value $emax(t)$.

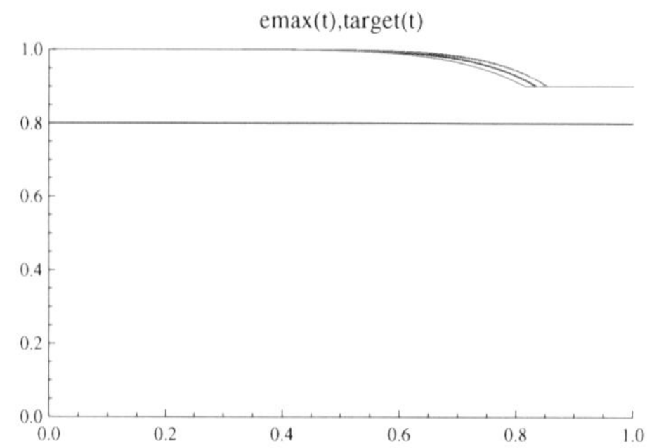

Fig. 6 Example of *target*(t) when *emax*(t) is set by expression 10

Figure 5 shows the acceptance probability space when the setting of *emax*(t) is assumed to be $\mu(t) = \frac{1}{10}t \quad d(t) = \frac{2}{5}t^2$, and the axis reaches time t and utility value o of the offer from the other agent.

Even if it is a small probability, the possibility of accepting by repeating actually remains. Therefore, when acceptance probability P is 0.1 or less, it becomes 0.

1.5 Correction in the Last Stage of Negotiation

Next, the correction in the last stage of the negotiation is shown. An excessive compromise was able to be controlled by introducing presumption concession degree *ratio*(t). However, the compromise stops when the other agent is proposing the mutual agreement candidate offer of a close utility value to its target and it doesn't change.

$$\mu(t) = \frac{4}{5}t \quad d(t) = 0 \tag{10}$$

Figure 6 is an example of *target*(t) when *emax*(t) is set by expression (10). The following expressions are given as a correction when the other agent is presenting the mutual agreement candidate offer of a close utility value to own target.

$$\gamma(t) = -300t + 400 \tag{11}$$

$$\delta(t) = target(t) - emax(t) \tag{12}$$

$$\varepsilon(t) = \begin{cases} \frac{1}{\delta(t)^2} & \text{if } \frac{1}{\delta(t)^2} < \gamma(t) \\ \gamma(t) & \text{otherwise} \end{cases} \tag{13}$$

$$\eta(t) = \frac{\delta(t) * \varepsilon(t)}{\gamma(t)} \tag{14}$$

$$target_2(t) = \begin{cases} target(t) - \eta(t) & \text{if } target(t) > emax(t) \\ emax(t) & \text{otherwise} \end{cases} \tag{15}$$

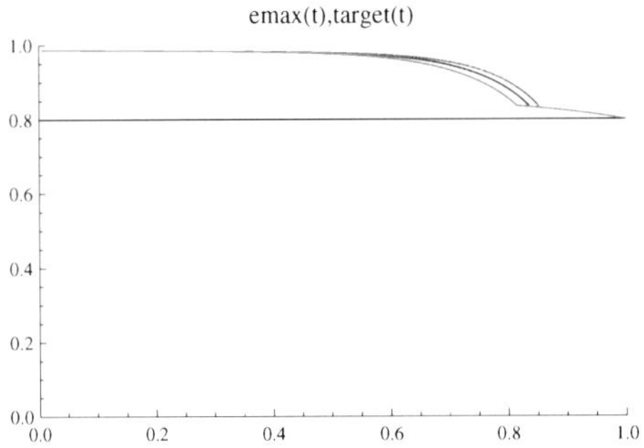

Fig. 7 Example of applying correction function to figure 6

$\gamma(t)$ is a function that adjusts the approach of the correction in the last stage of the negotiation. $\delta(t)$ shows the difference between targeted value $target(t)$ and estimated maximum value $emax(t)$. It limits it by $\gamma(t)$ though $\varepsilon(t)$ is in inverse proportion to the square of $\delta(t)$. $\eta(t)$ is a function that decides which extent to correct by using $\gamma(t)$,$\delta(t)$, and $\varepsilon(t)$ When $target(t)$ when coefficient $beta(t)$ is used is larger than $emax(t)$ correction is added to $target(t)$, and it is assumed a final target according to $\eta(t)$ function. It substitutes it by $emax(t)$ when $target(t)$ is smaller than $emax(t)$. This is assumed to be expression (15) as a final standard.

A constant approach is done when the distance with the other party is short when the correction is introduced. Therefore, the approach doesn't stop as shown in figure 6. It acts as shown in agent chart 7.

1.6 Strategy Corresponding to Discount Factor

$$DiscountTarget(t) = target(t) * DiscountRate(t) \tag{16}$$

$$DiscountTarget(t) * (1 - \text{BOU}) + target(t) * \text{BOU} \tag{17}$$

This is a strategy corresponding to Discount Factor. First strategy, Target is corrected by using the discount rate.(16) But, It decreases too much in this correction. Then, it adjusts it using the Best Offered opponent bid(BOU).(17) As a result, it is prevented from decreasing too much.

Printed by Publishers' Graphics LLC
MRO20130413.15.33.3